T0176739

Univariate, Bivariate, and Multivariate Statistics Using R

Univariate, Bivariate, and Multivariate Statistics Using R

Quantitative Tools for Data Analysis and Data Science

Daniel J. Denis

Registered Office
John Wiley & Sons, Inc., 111 River Street, Hoboken, NJ 07030, USA

Editorial Office
111 River Street, Hoboken, NJ 07030, USA

For details of our global editorial offices, customer services, and more information about Wiley products visit us at www.wiley.com.

Wiley also publishes its books in a variety of electronic formats and by print-on-demand. Some content that appears in standard print versions of this book may not be available in other formats.

Library of Congress Cataloging-in-Publication Data applied for

ISBN: 9781119549932

Cover Design: Wiley
Cover Image: © whiteMocca/Shutterstock

Set in 9.5/12.5pt STIXTwoText by SPi Global, Pondicherry, India

Printed in the United States of America

V10017991_031320

To Kaiser

Contents

Preface

In many departments and programs, a 16-week course in **applied statistics** or **data science,** even at the graduate level, is all that is allotted (unfortunately) for applied science majors. These include courses in psychology, sociology, education, chemistry, forestry, business, and possibly biology. The student must assimilate topics from so-called "elementary" statistics to much more advanced multivariate methods in a 3-month period. For such programs, students need a good introduction to the most commonly used fundamental techniques and a way to implement these techniques in software. For such courses and programs, it is hoped that the current book will fill this need.

This Book's Objective

This book is an elementary introduction to univariate through to multivariate applied statistics and data science featuring the use of R software. The book surveys many of the most common and classical statistical methods used in the applied social and natural sciences. What this book is not, is a deep, theoretical work on statistics, or an in-depth manual of every computational possibility in R or advanced visualization. Instead, the primary goal of the book can be summarized as follows:

This book emphasizes getting many common results quickly using the most popular functions in R, while at the same time introducing to the reader concepts relevant to data analysis and applied statistics. In this spirit, the book can be used as a general introduction or elementary primer to applied univariate through to multivariate statistics and data science using R, focusing on the most core and fundamental techniques used by most social and natural scientists.

The book is designed to be used in **upper division undergraduate** through to **graduate** courses in a wide range of disciplines, from behavioral and social science courses to any courses requiring a data-analytic and computational component. It is primarily a "how to" book, in most cases providing only the most essential statistical theory needed to implement and interpret a variety of univariate and multivariate methods. Computationally, the book simply "gets you started," by surveying a few of the more common ways of obtaining essential output. The book will likely not be of value to those users who are already familiar with the essentials both in theory and computation, and wish to extend that knowledge to a more sophisticated and computational foundation. For those users, more advanced sources, both in theory and computation, are recommended.

While the book can in theory be used without much prior exposure to statistics, an introductory course in statistics at the undergraduate level at some point in the reader's past is preferred. Most of what is learned in a first course in statistics is usually quickly forgotten and concepts never truly mastered on a deeper level. However, a prior introductory course serves to provide at least some exposure and initiation to many of the concepts discussed in this book so that they do not feel completely "new" to the reader. Statistical learning is often accomplished by **successive approximations** and **iterations**, and often prior familiarity with concepts is revealed as the material is learned again "for the first time." What you may grasp today is often a product of prior experience having a long history, even if the mastery of a concept today can feel entirely "sudden." The experience of "Oh, now I get it!" is usually determined by a much longer trail than we may at first realize. What suddenly makes sense now may have in one way or another been lurking below the limen of awareness for some time.

The book you hold in your hands is similar in its approach to a prior book published by the author featuring applied statistics, that one using SPSS, **SPSS Data Analysis for Univariate, Bivariate, and Multivariate Statistics (2019)**, also with Wiley. Both this book on R and the earlier one on SPSS are special cases of a wider more thorough text also published by the author, **Applied Univariate, Bivariate, and Multivariate Statistics (2016)**, also with Wiley (a second (and better) edition is set to be published in 2020). That book, however, while still applied, is a bit heavier on theory. Both the SPSS book and this one were written for those users who want a quicker and less theoretical introduction to these topics, and wish to get on **fast** with using the relevant software in applying these methods. Hence, both the SPSS book and this one are appropriate for courses, typically 16 weeks or shorter, that seek to include the data-analytic component without getting too bogged down into theoretical discussion or development. Instructors can select the book of their choice, SPSS or R, depending on the software they prefer. What sets this book apart from many other books at this level is **the explicit explanation provided in interpreting software output**. In most places, output is explained as much as possible, or quality references are provided where further reading may be required for a deeper understanding.

A few features that are designed to make the book useful, enjoyable to read, and suitable as an advanced undergraduate or beginning graduate textbook are as follows:

- **Bullet points** are used quite liberally to annotate and summarize different features of the output that are most relevant. What is not most relevant is generally not discussed, or further references are provided.

- "**Don't Forget!**"

  Brief "don't forget" summaries of the most relevant information appear throughout the book. These are designed to help reinforce the most pertinent information to facilitate mastery of it.

- **Exercises** appear at the end of each chapter. These include both conceptual open-ended questions (great for discussing and exploring concepts) as well as exercises using R.

What Does It Mean to "Know" R?

A General Comment on Software Knowledge versus Statistical Knowledge

R users range from those who are able to use the software with varying success to analyze data in support of their scientific pursuits, to those who are proficient experts at the software and its underlying **language**. Most of the readers who are using and learning from this book will fall into the former category. Computing, at the level of this book, isn't "rocket science," and you do not have to be an "R expert" to get something to work to analyze your data. You do not have to know the intricacies of the R language, the formalities of the language, etc., in order to use R in this fashion. All you really need is the ability to try different things and a whole lot of patience for figuring things out and **debugging** code. Though we do demonstrate quite a few of the functionalities available in R, we more or less stick to relatively easy-to-use functions that you can apply immediately to get results quickly.

Beyond the R capabilities in elementary introductions such as this one are veritable true **experts** in R who can literally generate programs on the fly (well, they too proceed by trial and error, and debug constantly too). These are the ones who truly interact with the software's language on a deeper level. Hence, in response to a question of "How do I do such-and-such in R?" these experts, even if they don't know the immediate answer, can often "figure it out," not through looking up code, but through understanding how R functions, and programming code more or less "on the spot" to make it work. We mention this simply so that you are aware of the level at which we are instructing on using R in this book. Beyond this basic level is a whole new world of programming possibilities where your specialization becomes not one of the science you are practicing, but rather of **computer programming**. Indeed, many of those who contribute packages to R are programming specialists who understand computing languages in general (not only R) at a very deep level. For most common scientific applications of R such as those featured in this book, you definitely do not need to know the software at anywhere near that level. What you need to know is how to make use of R to meet your data-analytic needs.

However, the same cannot be true of the **statistics** you are using R for. Understanding the statistics is, not surprisingly, the more difficult part. You can't "look up" statistical **understanding** as you can R code. Let me give you an example of what I mean. Consider the following exchange between students:

Student 1: "Do you know how to create a 3-D scatterplot in R?"
Student 2: "No, but I can look it up and figure it out using what I know about R, and keep trying until it works."

The above is categorically different from the following exchange:

Student 1: "Do you understand the difference between a p-value and an effect size?"
Student 2: "No, but I can look it up and figure it out and keep trying until I do."

The first scenario involves already understanding the nature of multivariable scatterplots and simply "figuring out" how to get one in R. The second scenario requires more understanding of applied statistics that develops over time, experience, and study. That is, it isn't simply a matter of "digging something up" on the internet as it was for the first example. The second scenario involves understanding the principles at play that develop via study and contemplation. In the age of software, knowing how to generate an ANOVA through software, for example, does not imply knowledge of how ANOVA works on a deeper level, and this distinction should be kept in mind. As I like to say to prospective Ph.D. candidates, assuming your dissertation is on a scientific topic, nobody at your defense meeting is going to ask you (or "test you on") how you computed something in R. However, someone will likely ask you to explain the meaning behind what you computed and how it applies to your research question. Now had your dissertation been titled "Computational Methods in R," then questions about how you communicated with R to obtain this or that output would have been fair and relevant. For most readers of this book, however, their primary topic is their chosen science, not the software.

It is important to remain aware of the distinction between statistical knowledge and software knowledge. In this book, software knowledge can be "dug up" as needed, whereas statistical knowledge may require a bit more thinking and deliberating. Being able to generate a result in software does not equate to understanding the concepts behind what you have computed.

Intended Audience and Advice for Instructors

As mentioned, the book is suitable for upper-division undergraduate or beginning graduate courses in applied statistics in the applied sciences and related areas. Because it isn't merely an R programming manual, the book will be well-suited for applied statistics courses that feature or use R as one of its software options. Depending on the nature of the course and the goals of the instructor, the book can be used either as a **primary text** or as a **supplement** to a more theoretical book, relying on the current work to provide guidance using R. Experienced instructors may also choose to develop the concepts mentioned in the book at a much deeper level via classroom notes, etc., while using the book to help guide the course. Because definitions can sometimes appear elusive without examples of their use, many concepts in this book are introduced or reviewed in the **context** of how they are used. This facilitates for a student the meaning behind the concept, rather than memorizing imperfect definitions which are never perfectly precise accounts of the underlying concept. Text in **bold** is used to provide emphasis on the word, concept, or sentence.

I hope you enjoy this book as a useful introduction to the world of introductory to advanced statistics using R. Thank you to my Editor, Mindy Okura-Marszycki, and all at Wiley who made this book possible, as well as students, colleagues, and others who have in one way or another influenced my own professional development. Please contact me at daniel.denis@umontana.edu or email@datapsyc.com with any comments or corrections. For data files and errata, please visit www.datapsyc.com.

Daniel J. Denis
January, 2020

1

Introduction to Applied Statistics

LEARNING OBJECTIVES

- Understand the logic of statistical inference, the purpose of statistical modeling, and where statistical inference fits in the era of "Big Data."
- Understand how statistical modeling is used in scientific pursuits.
- Understand the nature of the p-value, the differences between p-values and effect sizes, and why these differences are vital to understand when interpreting scientific evidence.
- Distinguish between type I and type II errors.
- Distinguish between point estimates and confidence intervals.
- Understand the nature of continuous versus discrete variables.
- Understand the ideas behind statistical power and how they relate to p-values.

The purpose of this chapter is to give a concise introduction to the world of **applied statistics** as they are generally used in scientific research. We start from the beginning, and build up what you need to know to understand the rest of the book. Further readings and recommendations are provided where warranted. It is hoped that this chapter will be of value not only to the novice armed with an introductory statistics course at some point in his or her past, but also to the more experienced reader who may have "gaps" in his or her knowledge and may find this chapter useful in unifying some principles that may have previously not been completely grasped. Thus, we launch into the book by introducing and revisiting some "big picture" items in applied statistics and discussing how they relate to scientific research. Understanding these elements is crucial in being able to appreciate how statistics are applied and used in science more generally.

Univariate, Bivariate, and Multivariate Statistics Using R: Quantitative Tools for Data Analysis and Data Science, First Edition. Daniel J. Denis.

1.1 The Nature of Statistics and Inference

The goal of scientific research can be said to learn something about **populations**, whether those populations consist of people, animals, weather patterns, stars in the sky, etc. Populations can range in size from very small to extremely large, and even infinite in size. For example, consider the population of Americans, which is very large, yet not as large as the population of stars in the sky, which can be said to be practically, even if not definitively, infinite. The population of coin flips on a coin can be regarded as an infinite population. Now, we often collect **subsets** of these larger populations to study, which we call **samples**, but you should always remember that the ultimate goal of scientific investigations is usually to learn something about populations, not samples.

So if the goal is to learn about populations, why do researchers bother with collecting much smaller samples? The answer is simple. Since populations can be extremely large, it can often be very expensive or even impossible to collect data on the entire population. However, even if we could, thanks to the contributions of **theoretical statistics**, primarily developed in the early twentieth century with the likes of R.A. Fisher and company, it may not be necessary to study the entire population in the first place, since with inferential statistics, one can study a sample, compute statistics on that sample, then make a quality inference or "educated guess" about the population. The majority of scientific research conducted on samples seeks to make a generalization of the sort:

If these facts are true about my sample, what can I say about them being true about the population from which my sample was drawn?

In a nutshell, this is what inferential statistics is all about – trying to make an educated guess at population parameters based on characteristics studied in a sample. If all researchers had access to population data, statistical inference would not exist, and though many of the topics in this book would still have their place, others would definitely not have been invented.

The primary goal of most scientific research is to learn something about populations, not samples. We study samples mostly because our populations are too large or impractical to study. However, the ultimate goal is usually to learn something about populations. If most of our populations were small enough to study completely, most of the fields of inferential statistics would likely not exist.

There are essentially two kinds of statistics: **descriptive** and **inferential**. Descriptive statistics are used to, not surprisingly, describe something about a sample or a population. The nature of the description is **numerical**, in that a formula (i.e. some numerical measure) is being applied to some sample or population to describe a characteristic of that sample or population. The goal of inferential statistics is to obtain an estimate of a **parameter** using a **statistic**, and then to assess the goodness of that estimate. For example, we compute a sample mean, then use that **estimator** to estimate a population parameter. The sample mean is a descriptive statistic because it describes the sample, and also an inferential statistic because we are using it to estimate the population mean.

1.2 A Motivating Example

Understanding applied statistics is best through easy-to-understand research examples. As you work through the book, try to see if you can apply the concepts to research examples of your own liking and interest, as it is a very powerful way to master the concepts. So if your area of interest is studying the relationship between anxiety and depression for instance, keep asking yourself as you work through the book how the techniques presented might help you in solving problems in that area of investigation.

Suppose you have a theory that a medication is useful in reducing high blood pressure. The most ideal, yet not very pragmatic, course of action would be to give the medication to all adults suffering from high blood pressure and to assess the proportion of those taking the drug that experience an adequate reduction in blood pressure relative to a control condition. That way, you will be able to basically tap into the entire population of American adults who suffer from high blood pressure, and get an accurate assessment as to whether your medication is effective. Of course, you can't do that. Not only would you require the informed consent of all Americans to participate in your study (many would refuse to participate), but it would also be literally impossible to recruit all Americans suffering from high blood pressure to participate in your study. And such a project would be terribly impractical, expensive, and would basically take forever to complete. What is more, however, thanks to inferential statistics, you do not need to do this, and can get a pretty good estimate of the effectiveness of the medication by intelligently selecting a sample of adults with high blood pressure, treating them with the medication, and then comparing their results to one of a control group, where the **control group** in this case is the group not receiving treatment. This would give you a good idea of the effectiveness of the drug on the sample. Then, you can infer that result to the population.

For instance, if you find a 20% decrease in high blood pressure in your sample relative to a control group, the relevant question you are then interested in asking is:

What is the probability of finding a result like this in my sample if the true reduction in symptomology in the population is actually equal to 0?

To understand what this question means, suppose instead that you had found a 0.0000001% decrease in high blood pressure in your sample when compared to a control group. Given this sample result, would you be willing to wager that it corresponds to an effect in the population from which these data were drawn? Probably not. That is, **the probability of obtaining such a small sample difference between the experimental group and the control group if the true population parameter is equal to 0 is probably still pretty high**. In other words, even though you found a small deviation in the sample, it probably isn't enough to conclude a true effect in the population. Why not? Because even on an intuitive basis, let alone a theoretically rigorous one, we expect a bit of **random variation** from sample to sample. That is, even if the true effect in the population were equal to 0, we'd still likely expect our sample to generate a result slightly different from 0. So if we did get such a small effect in our sample, it probably wouldn't be enough to conclude an effect in the population from which the data were drawn. That's it. If you understand that, you understand statistical inference, at least on an intuitive level. Observe a result in a sample, then ask the question as to whether what you're seeing in the sample is enough to conclude something is occurring in the population, or whether what you're seeing in the sample is best explained by just random variation or "chance."

1.3   What About "Big Data"?

Due to the high volumes of data able to be collected in industries such as in business, marketing, healthcare, the internet, and others, some data sets can be extremely large, and in some cases be near representations of actual populations. Recently, the **Big Data** "craze" has consumed many industries. For example, **Amazon.com** may claim to "know you" based on your purchase preferences and product history. Data of this kind occurs in overwhelming sizes. Healthcare is also an industry that, thanks to computers and essentially limitless ways of collecting and storing data, can maintain huge files on patient history and more easily and readily obtain patient records via computer networks and systems. Weather mapping also uses large data bases to provide the most accurate predictions possible of future weather patterns and hurricanes. The banking industry can rely on massive amounts of data to predict credit card fraud based on the number of times you use your credit card in succession or the number of times you incorrectly attempt to log into your online bank account. Indeed, the number of ways data can be collected and used in such applications is probably endless. Extremely large data sets can overwhelm memory capacities on typical desktop or laptop computers, and often requires **distributed computing** on several machines and networks in order to engineer and manage such large repositories of data. When dealing with such large data sets, programming and data management skills using software such as **Python** is often preferred. Such general-purpose programs are typically well-suited for data processing, and can also be used for statistical analyses. Programs such as **Hadoop** are also useful for distributing data sets to numerous computing machines. If you are working with extremely large data sets, you may also wish to familiarize yourself with R's **data.table**.

However, just because we can invent a new word for something, it does not necessarily imply the new word always represents something entirely novel. Marketers love new words to describe old things, because it draws in new customers (and students). Most of what is called "Big Data" is actually about technological and computer-networking and storage advance, and the **kinds of data** that are being collected, rather than the **nature of data** itself. True, the capability of collecting data and the ways of analyzing and visualizing data has definitely progressed in the past 20–30 years, what data **communicates**, and the limitations therein are still the same as they were before the rise of so-called "Big Data."

For example, while collecting larger data sets and being able to analyze more variables, "Big Data" does not in itself solve the **measurement problem**, that is, whether the data point provides the information you truly want to know. Yes, Big Data may be able to predict that if you bought a computer online you're more likely to next buy a mouse pad, but no algorithm can know **why** you bought either the computer or mouse pad. It can only predict, and will in many cases be wrong. Why are you getting a recommendation to purchase a new monitor after purchasing a new desktop computer? Because the algorithm has "learned" that many folks who buy computers next buy a monitor. However, the algorithm in your case may have gotten it wrong. You don't need a monitor, you only needed a computer, somewhat analogous to your car making an annoying "ding" sound when you take off your seat belt, "intelligent" enough to know you should be wearing it. However, not this time, because you're in a restricted driving area with no traffic going 5 miles per hour and will be exiting the car momentarily. Algorithms can only capture **patterns**, they cannot understand the uniqueness of each and every individual data point, and have no way of knowing whether the measurement of data is meaningful.

Hence, the **psychometric** issues remain. Does "Google know you"? Maybe, based on what kind of YouTube videos you watch, or how many hours per day you like to browse the internet. However, data of this kind, useful as it may be, is still relatively **shallow data**. It cannot know **why** you're watching this or that on YouTube, only that you are. Maybe an algorithm can predict what kind of car you will likely buy based on your past purchases and search history. But it cannot know **why** you are buying that car. The problem of **causation** (or at minimum, **directional flow**) still remains, if causation is to be taken seriously at all given the recent advances in quantum physics. In the applied behavioral sciences as well, the problems associated with self-report and validity of measurements still exist, regardless of the advances in software technology to store and analyze such information.

What is more, Big Data does not necessarily address the concerns of experimental sciences such as in biology, psychology, chemistry, and others. In these sciences, not only are large data sets unavailable, but they are also usually not at all practical to the experimental design. In these sciences, often researchers randomize cases to conditions, observe the influence of an experimental effect, and build a **mathematical model** that approximates how the data were generated. These are precisely performed experiments, and often participant recruitment is expensive. Evaluating whether a cancer treatment might be effective does not require Big Data. Rather, it requires an extremely and precisely performed experimental design on a cost-efficient sample, and the tools by which an inference can be made to the population from which the participants were drawn. That is, **the majority of scientific experiments require classical statistics and inference**.

Hence, though Big Data may represent an advance in computing technology and data management, at its core are still the elements of statistics, inference, experimental design, and the drawing of substantive conclusions from empirical observations. That is not to say there isn't anything "new" to Big Data, but it is to recognize that much of it is based on the same fundamental principles that existed before the "data craze," which includes many of the topics discussed in this book. **Data science** as well has its own terminology, though many of the terms mimic traditional data issues that have long existed. For instance, "tidying" a data set, which is popular language in data science, typically means ensuring it is entered and stored correctly, with rows generally indicating variable names and columns the objects under study. Transforming data, in data science terms, generally means selecting special features of data, for instance all those aged 50 or above, or all those with doctoral level education. It also includes constructing variables based on knowledge of other variables such as in $Y = X + Z$, where Y is the new variable, and X and Z are old variables used in its construction. When we **tidy** and **transform** data, we are said to be

data wrangling. Truth be told, "data wrangling" has been going on for centuries in one form or another and is nothing new, dating back to organizing data even when computers were not available (Newton and Faraday, for example, were professional data wranglers!)

1.4 Approach to Learning R

In this book, though we survey some of R's capabilities in the following chapter, many of the more common data analytic tasks and functions we present are in conjunction to the statistical procedures we study in the book. This has the advantage of not overwhelming the reader with a bunch of computational data management tasks all at once without appreciating the relevance of how and where they are commonly used. For example, creating a **factor variable** is a data management task, but it is most appreciated in the context of ANOVA models. Hence, we introduce many of the computational data management tasks when and as they are needed, so that you can immediately see how they are used. Of course, this book does not cover everything, and so when you need to accomplish a data management task, the internet or any of the other more popular data analysis or data science texts are your best go-to sources. No book or course can cover everything. Both are snippets of a wider project of life-learning (just as obtaining a PhD for example, is just the beginning of one's learning).

1.5 Statistical Modeling in a Nutshell

The basics of understanding statistical inference is no more complex than what we just described earlier regarding samples and populations. If it's that easy, you might ask, then why the rest of this book, and other books much more complex filled with formulas and long explanations? The reason is that researchers usually build what are called **statistical models** on samples of data, then wish to generalize these models to populations. However, this involves pretty much the same principles as we discussed above regarding inference, and as you'll learn to appreciate later, even inferring a difference between an experimental group and a control group was, strictly speaking, an example of a statistical model. In applied statistics, it's all about statistical models, and the job of an applied statistician is twofold:

1) Fit a model to sample data, and calculate how well that model fits the data.
2) Determine whether the sample model generalizes well to the population from which the sample model was built.

Let's consider a simple example based on a data set we will return to repeatedly over the course of this book. The following is a scatterplot of data on parent and child heights taken from Galton's data on families. For the parent height, an average of the two parents was used as the data point, and the child height is their height as grown adults (not as toddlers). A scatterplot of the data reveals the following plot (left):

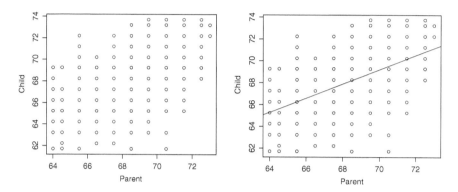

We can see that the data follow an approximate linear trend. We will perform what is called a **linear regression** on the data, which will fit in the swarm of data what is known as a **line of best fit**, or a **least-squares line** (right plot). We will study the details of how such a line is fit later in the book, but for now, it's enough for the point we are trying to make that such a line is fit using the following equation for a line that may already be familiar to you:

$$y = a + bx,$$

where y represents the (predicted) value for child, a is the intercept of the equation which represents where the line touches the ordinate axis, b is the slope of the line, that is, the degree to which the line is flat (slope of 0) or titled (for the current plot, the slope is positive), and x which is the value we input for parent. We haven't done anything "high-level," we simply fit a line to the swarm of data points that best accounts for the scatter. How that line is fit, and the equations used for it, will be the topic of a later chapter, but the point so far is that we fit a well-fitting line that seems, in some sense, to account for the scatter.

Of course, however, the line does not fit the swarm **perfectly**, and this is due to the fact that there is much scatter around our line. So, for a given value of parent, though our predicted value of child will be the value that falls on the line, we won't be right all the time with our prediction. That is, most often, we'll make an **error in prediction**, in that the actual value of child will often be higher or lower than the value on the line. Getting a grip on this error in prediction, that is, quantifying the degree to which our predictions are uncertain, is what statistical modeling is all about. Hence, to represent these errors in prediction, we append an "*e*" to our equation:

$$y = a + bx + e$$

Adding an error term *e* to our equation brings us into the territory of statistics. Drawing lines and other curves can be said to be an area of mathematics, more specifically, the kind of mathematics we all learned in high school. What makes statistics so much different from mathematics is the idea of **uncertainty** and how that uncertainty is measured and taken into account. A failure to understand this distinction is one reason why some students who do well in early college math courses sometimes struggle in their statistics course. They fail to understand that statistics isn't about plotting functions alone (such as a line). Rather, it's about plotting that function in the midst of a messy swarm of data, then trying to measure how well that function accounts for the data.

Finally, in line with what we said earlier about inference, we usually aren't that interested in this specific function (again, a line in this case). We are much more interested in the **population linear function** corresponding to population data. Notice the problem of inference hasn't changed simply because we are now considering a more elaborate model than the control versus experimental group situation in our first example. That is, we are still interested in conducting **inference**. For the linear equation, we will use *a* and *b* as estimates of the corresponding population parameters, α and β. Hence, we rewrite our regression model in terms of its population parameters,

$$y = \alpha + \beta x + \varepsilon,$$

where α is the population intercept, β is the population slope, and ε is the true error term. The linear regression model is the most popular model in statistics. However, it is not the only model (otherwise much of this book would not have had to be written!). Still, at their core, many (not all) of the other models featured in statistics are not that much different than the linear regression model. It may be difficult to see right now, but things like *t*-tests and analysis of variance models are actually **special cases of the wider linear regression model**. Other techniques such as **principal component analysis** and **cluster analysis** are different from the classic linear model, but still have similarities with other techniques as well. The key component to take away from this discussion thus far is that **many statistical models are very similar in nature, and if you aim to grasp the "big picture" of modeling, you will be able to incorporate virtually any statistical procedure into your current knowledge base**. Never learn statistical methods by the "correct procedure" approach for this or that occasion. Memorizing "which statistical method to use when" will not get you over the hump to understanding how statistical modeling works in general. Rather, attempt to see the unification behind the different procedures, and you will gain insights into modeling that you never could have dreamed, and you will be able to select the correct statistical method based on a higher understanding of the principles at play.

Many statistical models, at their core, are quite similar in nature. But to see this similarity, you need to forever try to appreciate the over-reaching "big picture" of statistical modeling. Then, you will be able to incorporate virtually any statistical method into that existing knowledge base. This will pay more dividends in the long run than memorizing which statistical method to apply for a given research situation.

1.6   Statistical Significance Testing and Error Rates

Classical hypothesis-testing in statistics is referred to as **null hypothesis significance testing**, or **NHST** for short. In such a hypothesis test, an investigator begins with putting forth what is called a **null hypothesis**, which is often a statement of **no difference** or **no relation** between variables. However, it should be emphasized that the phrase "null hypothesis" does not necessarily have to be a statement of "nothingness," as it is also quite possible to state a null that the population correlation, for example, is equal to a number other than zero. The idea behind the null hypothesis is that it is the hypothesis to be "nullified" or rejected if data contradict it. What does this mean? It comes back to our discussion of probability. You state a null hypothesis, then obtain data and determine the probability of such data (i.e. before the sample was drawn) in light of the null. Referring back to our earlier example, the null hypothesis was that the difference between the experimental and control group was equal to 0. The alternative hypothesis was that the difference is not equal to 0. If the results of the blood pressure medication come back and the sample difference is equal to say, 2% difference between the control and experimental groups, the relevant statistical question in terms of inference is the following:

What is the probability of observing a difference in the sample as large or larger as 2% if indeed in the population there is no difference between the control and experimental groups?

Notice that our statement of "no difference" here is not about the sample; it is about what might be going on in the population. As mentioned earlier, researchers really don't care all that much about samples; they are collecting them because they seek to know something about populations. If the probability of a 2% difference in the sample is high under the assumption of no difference between the control and experimental groups, then we would have no evidence to suggest a difference in the population. However, if the probability of a 2% difference in the sample is low under the assumption of the null hypothesis, then this would serve to cast doubt on the null, and possibly reject it if the obtained *p*-value is small enough.

How small a *p*-value is small enough? That should differ depending on how much risk we are willing to tolerate in making a wrong decision. Classic significance levels are 0.05 and 0.01, but we can easily imagine situations where a level of 0.10 or even higher may be warranted. The key point here is that if the probability of the data (i.e. the result you've obtained from your experiment or study) is small enough, it casts doubt on the null hypothesis, and leads to a rejection of it in favor of an alternative hypothesis. However, we must be careful here, because a rejection of the null does not "prove" the alternative hypothesis you may have in mind. All a rejection of the null does is point to a **statistical alternative**. The substantive research reason why the null hypothesis is being rejected is for you to know as a researcher based on how well of a controlled experiment you did or did not perform. If you did not perform a well-designed experiment that had strong controls in place, then concluding something scientific from a rejection of the null hypothesis is much more difficult. That is, it is quite easy to reject a null, as we will see, but the substantive reason for why the null was rejected is much harder to arrive at and requires good science to narrow down the true alternative hypothesis.

1.7 Simple Example of Inference Using a Coin

The principles of statistical inference can also be demonstrated and summarized with a simple coin experiment. It is not an exaggeration to say that if you understand this simple example, you will be at the open door of understanding the concepts of statistical inference in their entirety. It's very, very simple, do not make it complicated!

Suppose you hold in your hand a coin. You assume the coin is fair, meaning that the probability of heads is equal to 0.5 and the probability of tails is likewise equal to 0.5. This serves as your null hypothesis:

$$H_0 : p(H) = 0.5$$

The alternative hypothesis is that the coin is not fair, that is,

$$H_1 : p(H) \neq 0.5$$

Now, to determine the actual "truth" behind the coin for certain, we'd have to flip the coin an infinite number of times, that is, literally flip it forever to get the true ratio of heads to tails. However, since we live in the real world and cannot do this, we settle on obtaining a sample of coin flips, and make a decision about the fairness or unfairness of the coin based on this sample. Suppose we decide to flip the coin 100 times and obtain 50 heads and 50 tails. Would you reject the null? Absolutely not, since in the language of statistical inference, obtaining 50 heads out of 100 flips is exactly what you would have expected to obtain under the assumption that the null hypothesis was true. That is, the result 50/50 lines up exactly with **expectation**, which means the probability of obtaining 50 heads out of 100 flips under the assumption of the null is very high. It's actually as high as it can be for the problem. As we will see in a later chapter, for this particular problem, we can compute this probability using what is known as the **binomial distribution**. However, realize that you don't need a fancy stats book or computer to tell you the probability of obtaining 50 heads out of 100 flips on a presumed fair coin is high. It makes perfect intuitive sense that it should be high. So your conclusion would go something like this:

Since the probability of getting 50 heads out of 100 flips is very high, we have insufficient evidence to doubt the null hypothesis that the coin is fair, or, in other words, to doubt that the probability of heads is equal to 0.5.

Now, imagine instead of getting 50 heads, you got 100 heads. That is, all of your flips came out to be heads. What would be the probability of such a result under the assumption that the null hypothesis were true? If you are following the logic in all of this, you'd have to agree that getting 100 heads out of 100 flips is a very **unlikely result under the assumption that the null hypothesis is true**. Since we know that getting 50 heads is what we would expect to have seen, we know intuitively that getting 100 heads is an unlikely result under the assumption of the null hypothesis. We don't need a computer or statistics book or computation to tell us this! We have a gut feeling that such a result in our sample would lead us to **doubt the null hypothesis** since the obtained result is so extreme under it. That is, our conclusion would go something like this:

Since the probability of getting 100 heads out of 100 flips is so small, we have sufficient evidence to doubt the null hypothesis that the coin is fair. That is, we have evidence to doubt that the probability of heads on the coin is equal to 0.5. We therefore have evidence that the true probability of heads on the coin is different from 0.5.

What do you think the actual probability of the event is? We haven't computed it yet, but we can imagine it is very small, definitely smaller than a conventional level such as 0.05 or 0.01. When $p < 0.05$ or $p < 0.01$, for instance, we call the result **statistically significant**. Note carefully that this is all statistical significance means. **It means we have witnessed an event that is, by all accounts, relatively unlikely and unexpected under the assumed model we started out with. Statistical significance, as we will discuss more thoroughly very soon, does not necessarily imply anything scientifically meaningful or important occurred in the experiment or study.** Let's repeat this because it is one of the most important lessons of this entire book:

Statistical significance does not necessarily imply anything scientifically meaningful or important occurred in the experiment or study.

1.8 Statistics Is for Messy Situations

Notice in the above example with the coin, we argued that we didn't really need a statistics book or software to tell us these probabilities. In each case, we kind of already knew where these probabilities would land, at least approximately. The two scenarios with the coin featured extreme results. In one case, the sample evidence with the coin came up exactly in line with expectation, that is, the case where we obtained 50 heads. In the other case, the sample evidence with the coin came up extremely contrary to expectation under the null, that is, in the case of 100 heads. In each case, the decision on the null (i.e. whether to reject or not) was more or less a "slam dunk." In other words, we were able to conduct the statistical inference on an intuitive basis alone. We were able to assess the probability of the data given the null just by reasoning through it.

So where then do all the complexities of statistical theory come in? Formal statistical inference is most useful when the sample result is not extreme, that is, when the result is somewhat ambiguous that we're not quite sure at first glance whether to reject the null or not. For instance, suppose that instead of the case of 50 heads or 100 heads, we obtained 60 heads. We ask the question:

What is the probability of a result like this, of 60 heads out of 100 flips, under the assumption (null hypothesis) of a fair coin?

Notice that this result is much more difficult to answer. This is why we need statistical inference, because the empirical result is kind of "foggy" and it isn't immediately clear whether the coin is fair or not. We know 60 heads deviates from the expectation of 50 under the null, but we don't know whether that deviation in the sample is sufficient for us to reject the assumption of the coin being fair. That's where we need fancy statistical **sampling distributions** and computations to tell us the probability of the event, so that we can hedge our bet on the null hypothesis. **Sampling distributions are theoretical probability distributions of the given statistic**, based on an idealized sampling of an infinite number of such statistics. Hence, we can see that formal statistical inference is most useful when the results in the sample aren't clear cut and dry. **Statistical inference is most useful when results are messy**.

Statistical inference is most useful and needed when the empirical results from a study or experiment are not definitive in one direction or the other. That is, when the results of an experiment are rather difficult to make sense of (e.g. 60 heads out of 100 flips), that is when formal statistical inference is most useful. When the finding is extreme (e.g. 0 heads or 100 heads out of 100 flips), then the decision on the null hypothesis is usually already quite obvious.

1.9   Type I versus Type II Errors

When we reject or fail to reject the null hypothesis, keep in mind that we don't actually know if it's false or not. All we are doing is performing an **action** and making a **decision** on it. Just like you may be unsure whether buying a Subaru is a good decision or not, you go with the probabilities and hope for the best. The same goes with evaluating the goodness of a null hypothesis. If the computed probability is small, you may decide to reject the null hypothesis, but it may be that you're making a false rejection of it, and in reality, the null hypothesis is actually not false. For example, even in the extreme case of obtaining 100 heads on 100 flips, you'd have to agree that it is, at least on a theoretical level, entirely possible that the coin is still a fair coin and that obtaining the 100 heads was more or less a "fluke." That is, in rejecting the null because of the extreme data, you could be making an **inferential error** or a **wrong rejection**. These types of errors, false rejections of the null, are referred to as **type I errors**, and are equal to whatever you set your significance level to. So, if you set your significance level to 0.05, then your risk of committing a type I error is likewise equal to 0.05.

On the other hand, when you decide to retain the null hypothesis, you risk another type of error. If the coin came out 50 heads out of 100 flips, you wouldn't reject the null. However, it's still entirely possible that the coin is not fair, and that you just happened to witness a series of results that lined up exactly with expectation under the behavior of a fair coin. That is, your decision is to not reject the null, but it is entirely possible that you should have, and more trials on the coin would have eventually revealed its "unfairness" in the long run. A failure to reject a false null hypothesis is known as a **type II error**, meaning that you should have rejected the null because in reality the null was false. But you didn't based on the sample evidence.

Again, take note that specifying the error rates and what the significance level for any given experiment should be, should depend on the given situation and the experiment at hand. So, whereas setting the significance level at 0.05 might be suitable for one experiment, it may be wholly unsuitable for another, and a more conservative level may be more appropriate such as 0.01. However, as we require a more stringent significant level, we increase the probability of committing a type II error, so a balance and compromise of sorts must rule the day. Yes, you can set your significance level at 0.0000001 and virtually guarantee that you won't make any type I errors, but the problem is that with such a stringent level, you're virtually guaranteed to never reject any null hypotheses. Therefore, you'll inevitably miss out on false nulls along the way. Again, a compromise of sorts must rule the day, and though researchers often arbitrarily set a level of 0.05 for many statistical tests, that's more a matter of **blind convention** that it is good science. A careful consideration of the risks of making a type I versus type II error should be contemplated for each and every experiment one conducts, and while 0.05 is conventionally used in many situations, it should not be an automatic level chosen for each and every statistical test.

1.10   Point Estimates and Confidence Intervals

Recall that if most populations were either relatively small or at least manageable in size, we would have little need for inferential statistics. Recall that the reason why we collect samples in the first place is because accessing the entire population is usually not feasible, or at minimum, it requires much more resources and expenses than we are prepared to throw at the problem. Hence, we rely on inferential statistics computed on samples to give us an idea of what may be occurring in the population from which the sample was drawn.

When we compute a number on a sample, the number is called a **statistic**. When we use that statistic to estimate a parameter in the population, we are computing what is called a **point estimate** of that parameter. For example, if we wished to know the mean IQ of a particular population, it would usually be impractical to collect the entire population and compute a population mean on all of its members. So instead, we collect a sample, compute the mean $\bar{y}$ and use that as an **estimator** of the population mean μ. Using an arrow to denote that we are estimating, we can summarize the task as follows:

$$\bar{y} \to \mu,$$

which is taken to mean that "$\bar{y}$ estimates μ." While $\bar{y}$, on average, is a good estimator of μ, meaning that statistically, it possesses many of the qualities we desire in an estimator, for any given sample, **the probability that $\bar{y}$ will be actually exactly equal to μ is essentially equal to 0**. As an example, consider IQ in a sample measured to be $\bar{y} = 100.0758$. Do you really think this number could ever be exactly equal to the population mean μ? Probably not, since $\bar{y}$ and μ are both based on a continuous scale, and so theoretically, we could carry a number of decimal places on $\bar{y}$ (as many as we wish) and virtually guarantee that its value is not equal to the population parameter it is designed to estimate! That does not mean $\bar{y}$ is a poor estimator of μ, it simply means estimating μ is a tough job, and providing only a single number to estimate it may prove very challenging. Maybe we can do better.

If $\bar{y}$ is extremely unlikely to cover the population mean μ, intuition suggests that our chances of covering μ might increase if we built a **margin of error** around $\bar{y}$. For instance, instead of using $\bar{y}$, what if we constructed the following as an estimator:

$$\bar{y} \pm 1 \to \mu$$

That is, instead of $\bar{y}$ estimating μ, we have now built a range of values around $\bar{y}$ that might have a higher chance of covering the population mean. Note that we simply added and subtracted a value of 1 above. This was only for demonstration. We will define this range much more precisely shortly. The point for now is that an **interval estimator** will prove to have a higher chance at covering the population mean than will a single point estimator. We will refer to this interval estimator, that is, a range of values that has a certain chance of covering the population parameter, as a **confidence interval**. Hence, we can define a confidence interval as follows:

A range of values computed from a sample that has a particular probability of covering the population parameter.

The first thing to realize about confidence intervals is that they can be applied to virtually any estimation problem, whether it be estimating a mean, a proportion, a median, a mean difference, etc. In our example, for simplicity, we are only surveying the case of estimating a mean. However, you should know the concept is entirely general, and confidence intervals can be computed for a wide variety of problems.

Let's now take our intuition above, and define the confidence interval a bit more precisely. First, we had said that the interval has a designated probability of covering the parameter. However, what is this probability? While we can formally define confidence intervals for a whole range of probabilities, the most common ones are **90**, **95**, and **99%** intervals. For our purposes, we will stick with 95% intervals.

Now, let's consider how to build an interval, precisely. Again, for now, we focus on building one for the mean. The ± value we referred to will add or subtract a certain number of **standard errors** σ_M away from the mean (we will discuss standard errors more later). For instance, for a confidence interval in which we have knowledge of the population standard deviation σ, we define it as:

$$\bar{y} - 1.96\sigma_M < \mu < \bar{y} + 1.96\sigma_M$$

We interpret the above to mean that over all possible samples, the probability is 0.95 that the range between $\bar{y} - 1.96\sigma_M$ and $\bar{y} + 1.96\sigma_M$ will include the true mean, μ. The reason why we are using values of 1.96 on either side is simply because these are the **critical values** in a z distribution that defines the 95% confidence interval. The exact range of values we obtain for a given confidence interval will depend on the particular sample we obtained. That is, **it is the sample on which the confidence interval is built that is the random component, not the parameter**. The parameter μ is assumed to be fixed. Hence, on each sample we could theoretically collect from the population and compute a confidence interval, we will obtain different computed limits for the mean. What the confidence interval says, formally, is that **95% of those intervals computed will cover the population mean**. The following depicts possible confidence intervals computed in repeated samples when estimating a parameter:

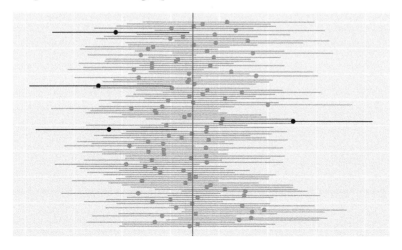

We can see that in this simulation, approximately 95% of intervals capture or cover the parameter of interest. A few of the intervals do not capture the mean. The percentage that do not cover the parameter is approximately 5%.

1.11 So What Can We Conclude from One Confidence Interval?

Some Philosophically Sticky Issues and a Brief Lament

If each interval we could theoretically compute in repeated sampling is virtually guaranteed to have a different range of values, then what could we possibly conclude from **one** confidence interval computed on one sample? That is, if we compute a confidence interval on one particular sample, and we obtain a range, say of 98–102, with the mean centered at 100, what can be concluded? Since 95% of intervals computed are thought to cover the population mean, we could conclude that if we were to sample repeatedly from this population and compute confidence intervals on each sample, then the **procedure** of computing intervals practically guarantees that 95% of these intervals would cover the mean. However, what knowledge of the actual parameter have we obtained? If you look at the above plot, it would seem that in our theoretical computations of intervals, we actually have very little knowledge of where the true parameter is! However, many writers on the subject will be most satisfied with the following interpretation:

We have 95% confidence that the true population mean is between the values of 98 and 102.

The "95% confidence" part is meant to imply the idea that in 95% of intervals computed in repeated sampling, 95% of these intervals will capture the population mean. This interpretation is what is usually considered to be the most "correct" way of interpreting a confidence interval.

If you're left unsatisfied with the above interpretation, you're not the only one. As suggested by the above figure, over repeated sampling, we would actually appear to have little if any certainty of where the true population mean lies, as the distribution of values away from the mean on each interval is quite sparse. So when we say we have 95% confidence that the true population mean is between the values of 98 and 102, do we really? You only computed a single interval, and what you wish to do as a researcher is to conclude something about the interval you computed.

For a slightly different interpretation, but one that we argue will likely be much more useful to you pragmatically as a researcher, we can adopt a definition similar to that of Crawley (2013), who computes a single interval on a sample with limits 24.76–25.74, and concludes the following:

> If we were to repeat the experiment, we could be 95% certain that the mean of the new sample would lie between 24.76 and 25.74 ... The confidence interval indicates the distribution of these hypothetical repeated estimates of the parameter. (pp. 122, 753)

This interpretation lends more meaning to the particular interval we computed on our sample and are using in our research, rather than visualizing a whole slew of confidence intervals providing different estimates in repeated sampling, all with their own respective margins of error. That is, to make the confidence interval most meaningful in a scientific context, the values you obtained are **plausible values for the parameter**, and this "plausibility" is conceived through repeated sampling from the population and obtaining sample estimates that lie within the bounds you computed. That is, in repeated sampling, you can expect the value of the parameter to lie within the bounds you computed.

1.12 Variable Types

In mathematics, a **variable** can be classified crudely into one of two types, **continuous** versus **discrete**. Mathematicians have developed extremely formal definitions of continuity which are well beyond the scope of the current book. In mathematics, formality is the name of the game, and concepts and terms are defined extremely precisely. With regard to continuity, the so-called **epsilon-delta** definition is often given in calculus or real analysis (which studies the theoretical basis of calculus) textbooks:

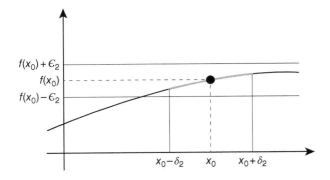

In the plot, we say that continuity exists at the given point $f(x_0)$ if, crudely (and without invoking limits), for a small change on the *x*-axis, either above or below x_0 (i.e. $x_0 + \delta_2$ or $x_0 - \delta_2$) we have an equally permissible small change on the *y*-axis (i.e. $f(x_0) + \varepsilon_2$ or $f(x_0) - \varepsilon_2$), right down to extremely small-as-we-like changes in delta δ_2 and epsilon ε_2. Hence, on a mathematical level, continuity is very, very **sharply** defined, and informally, invokes a "zeroing-in" process on the given point. For a more precise definition, see any introduction to calculus textbook.

In the world of applied science, however, and for our purposes, an informal definition of continuity will suffice, and will actually be more useful than the rigorous mathematical definition. Intuitively, continuity is best described by **drawing a line or other curve on a piece of paper and not once lifting the pencil from the page**. That is, what makes the curve continuous is that it has no **breaks** or **gaps** in it. If that line represents the range of possible scores on a variable, then the variable is considered to be continuous.

Now, contrast with the case where gaps or breaks do exist. That is, draw once more the curve, but at some point raise your pencil for a moment then lower it to continue drawing the curve. That "gap" you created had the effect of making the curve **discontinuous**, also generally known as **discrete**. The following figure depicts the difference between continuity and discreteness of a function in mathematics:

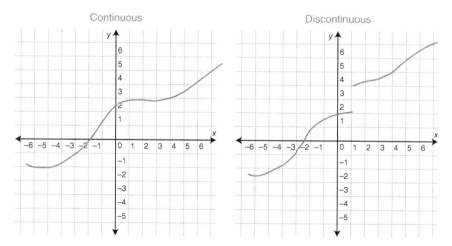

Now, our definition of continuity, informal as it may be, was still born from formal mathematics, even if we didn't define it as such. When we bring this definition down from the ideal ivory tower of mathematics into the real world of messy empirical variables and research, we need to somehow determine which variables are continuous and which are discrete. Surprisingly, whereas in mathematics the distinction is quite clear, with research variables, the distinction at times can be quite "fuzzy."

Let's consider an example. Consider the variable of height. Theoretically at least, it seems reasonable to think that we could measure height to any decimal point. That is, I could take your height and essentially record an infinite number of decimal places in my recording of it. In this way, height would be considered a continuous variable because any value of the variable is theoretically possible. In that sense, the difference between any two possible values of height goes to zero as we carry more and more decimal points. This idea is basically analogous to our formal mathematical definition of continuity. Now, contrast this to the variable of survival, where the two possibilities are "yes" versus "no" in an airplane crash. The variable of survival here would be considered "discontinuous" or "discrete" since one either survived the plane crash or did not, there is a "gap" or "break" between values of the variable. This idea is basically analogous to the mathematical definition of discreteness, because there is a "space" or "distance" between the two values of the variable.

It doesn't take long to realize that virtually all variables we actually measure in research are represented on a discontinuous scale, and that true continuity cannot realistically exist on real research variables once they are actually measured. However, often times it's useful for us to assume it exists for a variable on a theoretical level, such that the measurements could be assumed to have theoretically arisen from a continuous distribution.

When we measure height, for instance, we may assume it to be continuous in nature, though when we record height, we are recording "chunks" of it to put into our data file. Hence, if we really had to make hard conclusions about these distinctions, then we would have to admit that **true continuity cannot realistically exist with real research variables**. However, such a conclusion is not practical, and research is all about stepping down from the top of ivory perfection and accepting approximations. Pragmatically, the number of hairs on one's head may in theory be discrete, but since there are usually so many hairs, we often just assume the variable is continuous and compute statistical measures based on such an assumption. Hence, continuity is something scientists and data analysts have to simply assume for many of their variables and distributions. Such assumptions allow them to approximate their data using continuous distributions and curves. In other cases, however, variables are clearly discrete in nature, and even pretending or assuming them to be continuous is a stretch.

Why is this discussion of continuity relevant? It is relevant because as we consider a variety of statistical models in this book, one question that will keep popping up is whether the variables subjected to the analysis are best considered continuous or discrete. Some models only allow for more or less continuous variables, whereas others are more flexible. Often, the exact type of statistical model you adopt will depend a large part on whether variables are best considered continuous or discrete.

> Mathematically, continuity is defined very rigorously and precisely. More pragmatically, continuity is defined as drawing a line on a piece of paper and not lifting one's pencil at any point when drawing the line. In applied research, a variable is usually assumed to be continuous if we have reason to suspect its underlying distribution is continuous, or, there are a sufficient number of values of the variable that its distribution more or less approximates a continuous curve. Deciding whether a variable should be treated as continuous versus discrete is an important step for a researcher to make in selecting a statistical model.

1.13   Sample Size, Statistical Power, and Statistical Significance

How They Relate to One Another

If you are to interpret inferential statistical evidence, the following is by far the most important thing you need to know. In fact, it's so important, that I would go as far as to say if you're not clear on it, or have at best a fuzzy understanding of it, then you need to **perfect** your understanding of it before interpreting **any** statistical data that make inferences on population parameters. Rather than a long, drawn out theoretical argument, we immediately cut to the chase and summarize the issue with something you must always keep in mind when you interpret significance tests:

Given even the smallest of effect size in the sample, with large enough of a sample size, a rejection of the null hypothesis is assured for whatever test statistic we are computing.

That's it! That's the big one. Do not interpret any research papers using statistical inference until you clearly understand what the above means. And in the following couple of pages, I'm going to tell you exactly what it means and why it is so important.

First, let's consider what **effect size** means. Effect size is a fairly general term that is kicked around research circles and means slightly different things in different contexts, but ultimately you can think of it as a measure of the extent to which something "happened" in the sample you are studying. For instance, did the pill work and reduce symptomology in headaches? Was the medication effective in prolonging survival? Were letters of recommendation found predictive of graduate school success? If so, what was the explained variance? These are all examples of effect size type statistics. They tell you what happened in your sample, and whether something of scientific interest occurred. They are generally measures of **distance** or **association**. Not surprisingly, researchers should be consumed with finding impressive effects. Trying to find large effect sizes should be their primary goal.

The problem, however, at least historically in many of the sciences that rely on statistical inference to draw conclusions from data, is that it is entirely possible to obtain the infamous "$p < 0.05$" **even under circumstances where the effect size measure is exceedingly small**. How does this occur? It can occur in a few ways, but by far the most common way is when **sample size is large**. For instance, a mean difference in the sample of a control group versus an experimental group, regardless of how small that difference may be, can be made statistically significant if sample size is large enough. So, suppose you find a mean difference in the sample of 0.00001 between wellness scores of those treated with cognitive versus behavioral therapy in some experiment. You'd have to agree the difference is negligible, it's trivial. It's extremely small as to almost be invisible. However, with a large enough sample size, one can attain $p < 0.05$ and reject the null hypothesis of no difference in the population from which the data were drawn! And while this fact is not alarming on a theoretical level (i.e. it makes perfect sense to theoreticians), what is alarming is when scientists draw **substantive conclusions** of a "real" difference occurring based on p-values alone. Historically, scientists, both social and natural, have unfortunately too often equated statements of "$p < 0.05$" with a scientific effect, a sense of "success" for their experiment. But as noted, achieving statistical significance alone is insufficient evidence of such a scientific effect, as we will now demonstrate with a very simple example.

1.14   How "$p < 0.05$" Happens

The following brief discussion and demonstration of significance testing is adapted from Denis (2016, pp. 155–164), in which a much more detailed and elaborate discussion of the problem is given. Our discussion here is an abbreviated version of that more lengthy treatment and explanation.

The issues with misinterpreting significance tests are best discussed within the context of a very simple example. Suppose you are a psychologist who hypothesizes that implementing a new program for high school students will have the effect of improving achievement scores on a standardized test. For simplicity, suppose the mean achievement of all students in the United States on that test is equal to 100. That is, $\mu = 100$, where μ is the population mean. If your program is effective, you would expect to show a higher mean than 100 for those participating in your program. That is, your alternative hypothesis is that $\mu \neq 100$, or more specifically, $\mu > 100$.

To help evaluate your theory, you randomly sample 100 students from the state, and give them the 3-month program. Then, at the end of that period, they are administered a standardized achievement battery. Suppose the mean achievement of the sample comes out to be 101. That is, $\bar{y} = 101$. What do you think of the result? Forget any statistics for a moment. Do you think the result is impressive? Probably not, because the sample mean is only a single unit different from the population mean. That is, $101 - 100 = 1$. Is that an impressive result? Probably not, and if we were to conduct no statistical inference, we would probably simply conclude that we have insufficient evidence to think the program is effective. After all, if it were effective, we would probably expect a much greater difference, say of at least 5–10 points, maybe more. But the sample came out to be 101, which is very close to what we would have expected under the null hypothesis (recall the expectation under the null was 100).

You choose to evaluate your result for statistical significance anyway, and after computing the standard deviation in the sample to be equal to 10, you compute the ensuing t statistic:

$$t = \frac{\bar{y} - \mu_0}{s/\sqrt{n}} = \frac{101 - 100}{10/\sqrt{100}} = \frac{1}{1} = 1$$

The t statistic you obtained is equal to 1, and when evaluated on 99 degrees of freedom, is found, of course, to not be statistically significant. Hence, you fail to reject the null hypothesis. That is, you have insufficient evidence to reject $\mu = 100$. In this case, the significance test more or less accords with your scientific intuition, that the program is not effective, so no harm is done. That is, you didn't reject the null, you obtained a fairly high p-value, and all of this agreed with your sense of no scientific effect. In other words, the program doesn't work, and all is good, you can move on to another experiment or try something that might work.

Now, consider the exact same study as above, except that instead of testing a sample of 100 students, you had decided to test a **sample of 500**. However, the difference in means remained the same, as did the standard deviation. The new computation for t is:

$$t = \frac{\bar{y} - \mu_0}{s/\sqrt{n}} = \frac{101 - 100}{10/\sqrt{500}} = \frac{1}{0.45} = 2.22$$

Notice that *t* came out to be much larger this time, **even though the difference between means in the numerator remained exactly the same**. That is, note the difference between means is still equal to 1 point. This particular *t* of 2.22, if evaluated for statistical significance, would meet the 0.05 criterion. Hence, the null would be rejected, and you would be able to say you found evidence for a difference in means! Therefore, you would be completely honest in making a conclusive statement such as:

We reject the null hypothesis that the program mean *μ* is equal to 100, and conclude that it is not. The result is statistically significant at *p* < 0.05.

While the above is technically a correct statement, it is entirely open to misunderstanding if one does not first understand the make-up of a significance test and the influences of what makes a *p*-value small. As we have seen, an increasing sample size, all else equal (e.g. the distance in means remains the same) has the effect of virtually guaranteeing that at some point, the null hypothesis will be rejected. But of course, that's not why you did the study. You did the study to see if your program resulted in an appreciable increase in mean achievement. That you obtained *p* < 0.05 should not impress you that much. What should impress or "depress" you is the **effect size**, which as we will see, is not amenable to fluctuations in sample size as is the significance test.

Don't FORGET!

That you obtained a statistically significant result such as p < 0.05 does not necessarily imply that your experiment or study was a success! Without computing an effect size to go along with the significance test, whether or not your experiment demonstrated anything scientifically meaningful or relevant remains elusive.

1.15 Effect Size

Recall we said earlier that effect sizes, in essence, provide us with a measure of "what happened" in our study or experiment. They are a measure of **scientific effect**. And while the significance test is sensitive to sample size, effect sizes are less so. At minimum, they are not simply a function of sample size. As you collect a bigger and bigger sample, the effect size may or may not increase, while for the significance test, the value of *z* or *t* or *F* (or whatever inferential test statistic you are computing for the given study) is virtually guaranteed to increase in value.

For our example above, that of a mean difference, a common measure of effect size is to compute the difference in means and divide by the standard deviation. That is, we compute:

$$d = \left| \frac{\bar{y} - \mu_0}{\sigma} \right|,$$

where $\bar{y}$ is the mean of the sample, μ_0 is the mean under the null hypothesis, and σ is the population standard deviation, or an estimate of it in the form of s if we do not have the actual value. Technically, since $\bar{y}$ is serving as the estimate of μ, the numerator is more accurately represented as $\mu - \mu_0$. But noting $\bar{y}$ instead of μ, however, reminds us where we are obtaining this mean. The "$|\cdots|$" indicates to take the absolute value of the resulting number, since we are most interested in simply the **magnitude** of the difference between means, and not necessarily the sign. This measure of effect size is referred to as **Cohen's d** in the scientific literature. It is a **standardized difference between means**. The standardization takes place by dividing $\bar{y} - \mu_0$ by σ. For our data, the computation is:

$$d = \left| \frac{\bar{y} - \mu_0}{\sigma} \right| = \left| \frac{101 - 100}{10} \right| = 0.1$$

By all accounts, including those of Cohen's original guidelines for what constitutes a small (0.2), medium (0.5), or large (0.8) effect size, a value of 0.1 is quite small. Notice that the value for d is not "inflatable" or "deflatable" as is the p-value for an increase in sample size. That is, if we were to increase sample size, then σ in the denominator may increase or decrease. This is different from the standard error of the mean featured earlier for which an increase in sample size was almost surely going to make the standard error smaller, which meant the resulting test statistic will be larger, leading to more rejections of the null hypothesis. **Effect sizes do not automatically increase simply because one is using a larger sample size, whereas p-values typically get smaller as sample size increases.**

1.16 The Verdict on Significance Testing

Null hypothesis significance testing does have its place in science, but it must be used with an acute understanding of what p-values can and cannot provide for the scientist. They are still useful since they provide a measure of inferential support for the scientific finding, but too often historically, they have been mistakenly misinterpreted as automatically indicating that an effect of importance has been found in an experiment or study. As we have discussed, to make any statement of effect, one needs to compute effect size. Whereas *p*-values can indicate whether inferential support is warranted, which in itself is important, effect sizes give us a sense of the degree to which the independent variable accounts for variance in the dependent variable, or a sense of the magnitude of association found among variables in the experiment or study.

1.17   Training versus Test Data

Having just featured a discussion of sample size, and how it relates to *p*-values, we should at this point briefly touch upon another related matter, and that is the concept of **training** versus **test data**. In experimental studies especially, collecting large samples can be very time-consuming and expensive, and in many cases, are simply impossible or unlikely to happen. In these cases, achieving even lower limits of acceptable statistical power may be difficult. However, in other cases, large samples may be more accessible, especially in the age of "Big Data." In such cases, one may wish to separate the entire data set into a **training set** and a **test set**, fit the given model on the training set, then cross-validate it on the test set. Such a splitting procedure allows the researcher to obtain a bit more of an accurate assessment of model validation statistics. See Izenman (2008) for details.

In this book, we focus generally on fitting and interpreting models on **single** data sets. Many of the larger data sets we use are already a part of the base package in R. Most of the original and fictitious data sets featured in this book are relatively small, and the focus is on fitting them and interpreting them, not on evaluating them on test data. However, as you proceed further into modeling, if sample size allows, you may wish to cross-validate **all** of your models on partitions of the larger data frame as to obtain a more accurate estimate of model fit.

It is also important to recognize that the notion of cross-validation (and similar procedures), the **concept** at least, has always existed in mathematical modeling. But it has received renewed emphasis in the statistical and machine learning literature largely due to the advent, accessibility, and availability of larger data sets in some settings. In science, a related concept is that of **replication**. The model you obtain on one sample, especially if not hypothesis-driven, gains reputation only if it is tested on another sample. In this way, the experimenter wishes to evaluate how well the derived model fits new data, rather than the data on which it was generated. Though the matter of training versus test data is not an absolute equivalent parallel to the concept of replication in the sciences, it is nonetheless a useful, if not somewhat crude way for the experimenter to think about such issues in the more global context of scientific investigation. The moral of the story is clear; good researchers aim to replicate findings as to provide as accurate as possible measures of model validation, whether on partitions of the data that have been collected, or on future data. Only through replication can scientific phenomena, if present, be confirmed. Statistically, if you derive a model on one set of data, it is always a good idea to try to test and validate it on another as to obtain more valid estimates of model fit.

1.18 How to Get the Most Out of This Book

Having surveyed the important concepts of *p*-values and effect sizes, we are now ready to journey through the rest of this book and learn how to use R software to perform and interpret the most common of statistical methods. The following are some tips for getting the most out of the book:

- Focus on a **conceptual understanding** of what the featured statistical method accomplishes. Aim for a "big picture" bird's eye view, and as you require more details, you can come back to them later. Even so-called "experts" on a topic do not know everything. However, they do know where to "dig" to find the information they need. The key to knowledge is knowing where to dig, and especially when digging is required.
- Do not memorize either formulas or R code. Rather, seek to understand why the given formula or equation works the way it does, and the general format of the R code used. As you learn more statistics and coding, you will begin to see recurring themes and patterns in both. But there is no reason to try to "force" these upon yourself by trying to memorize things.
- Highlight and put asterisks next to things you do not understand, and return to these again later. Rarely are things perfectly understood upon a first reading. Do not allow yourself to get too "bogged down" on any particular element because you may not fully grasp it. Consuming technical information, whether in statistics or computing, takes a lot of time and iterations. Aim to deepen your understanding of things you believe you already know, and keep your mind open to sharpening that understanding as you learn more.

Lastly, remember that statistics and computing are not particularly easy subjects. Statistics, even at a relatively nonmathematical level, is still intrinsically difficult and I know of nobody who studies or uses statistics who doesn't find the subject very challenging. The less we know something, the more it feels like others must understand it so much more, but I can tell you that's not the case at all. Statistics, computing, and research are difficult subjects for virtually everyone, so have patience with yourself and aim to enjoy learning rather than getting bogged down and frustrated by it. The "baby steps" you are making in your learning are the same ones most others are making as well.

Exercises

1 Discuss the nature of statistical and scientific inference. When are inferential statistics needed to make scientific inferences? When are they not?

2 Distinguish between samples and populations, and how the advent of "Big Data" relates to the wider framework of statistical inference.

3 Discuss how "Big Data" does not necessarily address the measurement problem inherent in many sciences.

4 Describe what is meant by a statistical model, and why statistical modeling can be considered a special case of the wider framework of mathematical modeling.

5 Discuss the nature of null hypothesis significance testing, and the concepts of type I and type II errors when making a decision on the null hypothesis.

6 Explain the meaning of a confidence interval, and why a more pragmatic interpretation of it for a single research result may be useful.

7 Distinguish between continuous versus discrete variables, and why continuity in research variables likely can never truly exist, especially when juxtaposed to the definition of how true continuity is conceived in mathematics.

8 Define what is meant by an effect size in research. How is an effect size different from statistical significance?

9 Discuss and explain the determinants of "$p < 0.05$", and how statistical significance can be achieved with or without a large effect size.

10 Distinguish between training versus test data, and how these concepts relate to scientific replication in the applied sciences.

2

Introduction to R and Computational Statistics

LEARNING OBJECTIVES
• Understand in general how computing languages work and why R is especially good for statistics.
• How to install R onto your computer and start performing basic statistics with it immediately, including plotting graphs.
• How to get data into R and install packages.
• How to compute a variety of functions in R useful for data analysis and data science.
• Understand the fundamentals of matrix concepts used in statistical procedures featured in this book, and be able to perform these computations in R.

Due to the advance of computers in the last half century or so, the field of what is called **computational statistics** has arisen. This discipline can be considered a merger of statistics on the one hand, and computer science on the other. In the early 1900s, manually calculating statistical procedures such as ANOVA and regression was quite laborious, and advanced procedures such as factor analysis requiring complex matrix computations took, well, forever to complete. Today of course, one can subject data to one of the myriad of statistical programs available and obtain output in literally a matter of seconds. This can be considered not only as a boon to making use of available data and gaining insights from such data, but it can also sometimes foster not thinking through a problem sufficiently before generating analyses on it. Statistical computation is great, but the researcher needs to make sure he or she is sufficiently well-versed in the actual method before attempting to compute and interpret the method. As noted by Guttag (2013):

> A computer does two things, and two things only: it performs calculations and it remembers the results of those calculations. But it does those two things extremely well. (p. 1)

A discussion or debate as to whether or not a computer can actually **think** or not, is, not surprisingly, beyond the scope of the current book, but for the purposes of this book, it really doesn't matter. What matters for us is the ability to basically get the computer, and more specifically the R program, to produce what we want it to.

R is a **scripting language** designed specifically for statistical analysis and data management. It is a derivative of the S programming language, but is free and is continually updated by contributors through its numerous packages. R is different from many other packages such as SPSS or SAS, where traditionally the user submits a collection of scripted lines or mini programs, all at once, at which point the computer generates a bunch of output, some of it welcome and useful, the rest of it potentially superfluous. For many of the functions in R, it's a bit more akin to a conversation with the program than anything similar to a submission of a collection of lines. In that way, R is more similar to computing languages that computer scientists use every day, such as **Python** or **Java**, in that they are continually communicating with the computer and receiving feedback regarding what they are requesting it to accomplish. Basically, computing languages contain what are known as **primitives**, which are essentially expressive building blocks of the language. Much deeper are the 0s and 1s that make up the binary **machine code** of the language. From there, a higher level layer is produced which allows facility in computing what the language was designed for. For a statistical program, this higher layer will include, not surprisingly, many prepackaged statistical functions.

For instance, in computing an arithmetic mean, it would be a hassle if every time we wanted it we had to compute it the long way as in the following list of numbers:

```
> x <- c(0, 2, 5, 8, 9)
> sum(x)/length(x)
[1] 4.8
```

Above we have computed the arithmetic mean by summing observations on the vector (sum(x)), then dividing by the number of observations the sum is based, called the **length** of the vector (length(x)). Of course, any good statistical software will not require us to do this computation every time we wish to compute an average. Instead, software has built-in functions that perform these operations automatically, and much more quickly. In R, all we have to do to get the mean is compute:

```
> mean(x)
[1] 4.8
```

However, it should also be noted that at their core, computer languages can be reduced to what we referred to earlier, called **machine code**, which is a more fundamental level of the computing language. This refers to what is happening "behind the scenes" as the software program moves around bits of code. Indeed, if all computing languages were restricted to working in machine code, computing anything, even very simple things, would take forever! To get to a higher level in the language, software developers join bits of this machine code together to perform things automatically, such as compute the mean we just calculated. Then, when we request mean(x) for instance, the software already has the machine code pre-organized to perform this operation. That is, the computer already has the **algorithm** (yes, the computation of a simple average is an example of an algorithm!) set up so that we can simply make a **call** to that algorithm instead of moving binary pieces of information around to program it anew.

So why is any of this discussion relevant? It is relevant simply to better understand the course we are about to embark on, which is that of using a computing language (R) to generate statistical output that we then wish to interpret. Getting to know R means learning how to interact with it and getting it to do what you need it to do. It requires a tremendous amount of **trial and error**. Newcomers to programming are sometimes under the impression that "experts" in the area never make mistakes or receive error messages. How untrue! Believe me, even the best of programmers are continually getting error messages and debugging programs and scripts to make them work. **Getting error messages until something works is the art of programming**. The learning curve for scripting languages such as R can be a bit steep, and many times it can seem like the computer will never do what you need it to do, but with patience and perseverance, as is true in life in general, the lessons learned in each trial pays dividends down the road as you learn to develop mastery over the language (life).

The following image is a useful one that depicts the mix of skills one requires to effectively combine substantive knowledge of his or her field with the requisite computing skills:

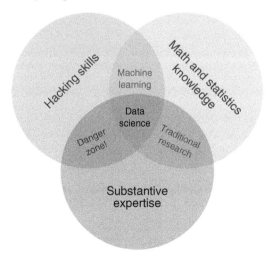

For most students, especially graduate students, the right mix is somewhere in the middle where we find "data science" and "traditional research." If you're going to learn and use R, you're going to end up doing a bit of "hacking" along the way. Note the danger zone, however, where having some substantive expertise but very little knowledge of mathematics and statistics sets one up to fail. Heed the following warning:

Never assume computing knowledge can replace substantive expertise and knowledge of mathematics and statistics. Obtaining output in any software program does not equate to understanding the output, or understanding whether what you produced is correct.

Having said all the above, and with the relevant caveats in place, let's get on with learning how to use R effectively to generate the statistical procedures we will need for the rest of the book. In what follows in this chapter, we survey only the basics. We'll pick up the rest as we work through a variety of statistical methods. In what follows, we survey some of the immediate things you need to understand to implement code and conduct statistical analyses in the rest of the book.

 Learning to compute with R requires a tremendous amount of trial and error, even at the levels of the highest expertise. The final goal is not to be able to compute something, but rather to understand in depth what you have computed, and for that, a solid knowledge base of statistics and mathematics is necessary.

2.1   How to Install R on Your Computer

R is an open-source software and as mentioned, is free. You can download a copy of it directly from the **R Project for Statistical Computing** website:

```
https://www.r-project.org
```

Once on the site, you will see on the left-hand panel near the top the word **Download**. Underneath that is the word **CRAN**. This will bring you to a site called **CRAN Mirrors**. All links contain identical content, so in theory at least, you can click on any mirror site you wish. However, it is usually best to select the mirror site more or less closest to your geographical location, as the download may be faster than if you selected a mirror site half-way around the world. We will select the first site under USA, University of California, Berkeley, CA:

```
https://cran.cnr.berkeley.edu/
```

Once you click on that link, it will bring you to a page with the title **The Comprehensive R Archive Network**. Under **Download and Install R**, select either R for Macs or R for Windows. For this example, we select **Download R for Windows**, which brings us to a new page titled **R for Windows**. Click on **install R for the first time**, then on the next page, **Download R 3.6.1 for Windows**. Save the **R-3.6.1-win.exe** file to a place on your hard drive, then once the program is done downloading, double-click on it to begin the installation process.

Once you are done installing R, open the program, and the following window will appear, which means R was installed correctly and you're ready to begin!

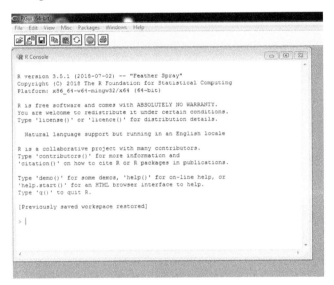

2.2   How to Do Basic Mathematics with R

Now that R has been successfully installed on your computer, it is time to get familiar with the software. One great thing about R is that you can think of it as an extremely powerful **calculator**, and begin doing things with it that you would otherwise use your pocket calculator for. Let's look at some examples of a few mathematical operations in R. Occasionally, # may appear in computer code such as R or Python, which simply means a comment. The symbol # is meant to tell the software that it is a comment and not to incorporate it into its computations; otherwise we would continually see error messages when we tried to label something with a comment.

The first thing you see in R is the prompt:

```
>
```

What this means is that R is ready for you to input something. For example, let's input the number 2:

```
> 2
[1]  2
```

After entering 2, hit **Enter** on your computer. R produces the number 2 in its output. This is the "stimulus-response" style that R functions in from the main interface, in that you enter something, and it responds with something (hopefully not an error message!). Let's enter a few more mathematical functions to demonstrate some of R's basic capabilities:

```
> 2 + 3
[1]  5

> 2/4
[1]  0.5
```

Be aware that R doesn't always recognize the way we often write mathematics. For instance, if we wanted to multiply 2 and 3 together, on paper we could write (2) (3) = 6. However, in R, this is what happens when we try this:

```
> (2)(3)
Error: attempt to apply non-function
```

In short, the program wasn't able to execute what we thought was a reasonable way of writing multiplication. Such is the process of learning a computing language, to learn what the given program "likes" versus what it does not like. For the multiplication problem, R likes the following much better:

```
> 2*3
[1]  6
```

R will obey the order of operations, as the following demonstrates:

```
> 2*3+5
[1]  11
```

Notice that R first multiplied 2 and 3, then added 5. This is the same thing we would do manually, so we can see that R knows "BEDMAS" (brackets, then exponents, then division/multiplication, then addition/subtraction). Moving on then, we can quickly see how R can handle more complicated series of arithmetical operations:

```
> ((2*3) + 6 - (2/3 - 11))/6^5
[1] 0.002872085
```

The last part of the above expression, 6^5 is exponentiation, that is, "6 raised to the exponent 5" or 6^5 (confirm for yourself that it equals 7776).

R can of course handle much more complicated mathematics than the above. For instance, it can compute **logarithms**, **exponentials**, **derivatives**, and **integrals**, to name just a few. For example, you may remember that the logarithmic and exponential functions are inverses of one another. We can demonstrate in R:

```
> log(10)
[1] 2.302585
> exp(2.302585)
[1] 9.999999
```

Within rounding error, we see $e^{2.302585}$ is equal to 10 (i.e. 9.999999). We can round the number to the nearest integer by using round(exp(2.302585)):

```
> round(exp(2.302585))
[1] 10
```

That is, the natural logarithm of a number to base e is the exponent to which we raise e to get that number. We will be using this inverse relationship between logs and exponents when we survey **logistic regression** later in the book. If you are familiar with derivatives (if you are not, no worries, you won't need to know them to understand this book), then you know that the derivative of the following function, $f(x) = 5x + 3$ is equal to 5. That is, $f'(x) = 5$. R can compute this for us:

```
> D(expression(5*x + 3),"x")
[1] 5
```

You can also control how many digits are printed in R. For example:

```
> print(pi, digits = 2)
[1] 3.1
> print(pi, digits = 10)
[1] 3.141592654
```

The above prints 10 digits to pi. Theoretically, pi may well go on almost forever, but a computer will never be able to capture "forever" and can only store a **finite**, as opposed to **infinite number of digits**. As an example of this, consider the following statement in which we evaluate whether the expression on the left is equal to the number on the right:

```
> 1/2 * 2 == 1
[1] TRUE
```

The above is obviously true, and R confirms that it is. However, 1/2 has a finite decimal point, in that it is equal to 0.5. Hence, 0.5 multiplied by 2 is easily seen to be equal to 1. However, had the expression read the following, we would have obtained a different answer:

```
> sqrt(2) ^ 2 == 2
[1] FALSE
```

Of course, we would have expected the statement to be true, since squaring a square root should eliminate that square root and simply return the number 2. However, the reason why the statement is false is because the number `sqrt(2)` is an approximation and does not have a finite decimal representation. Hence, the statement of equality cannot be true. Coding > `near(sqrt(2)^2, 2)` will generate a true statement (available in **dplyr** package) however, since the function `near()` allows for "close to" the number 2, instead of exactly equal to it as `==` requires. Don't be fooled either when R reports a finite decimal point to the number `sqrt(2)`. The following number does not have only 6 positions to the right of the decimal point. R is simply rounding the number to 6 spaces:

```
> sqrt(2)
[1] 1.414214
```

2.2.1 Combinations and Permutations

R is also useful for generating combinations and permutations. Recall that the number of **combinations** is the number of ways of choosing a number of objects out of a set of objects such that the **order of the objects does not matter**. For example, suppose we have 5 objects and wish to know the number of ways of choosing 2 of those objects if order doesn't matter:

```
> choose(5, 2)
[1] 10
```

That is, the number of ways of choosing 2 objects out of 5 if order does not matter is equal to 10.

The number of **permutations** is likewise the number of ways of choosing a certain number of objects out of a set of objects, but this time, **order does matter**. To compute the number of permutations of 5 choose 2, we calculate:

```
> factorial(5)/factorial(5-2)
[1] 20
```

This of course is the just the beginning of the mathematics one can do with R. You will learn more possibilities as you work through this book. Since the primary focus of the book is in producing statistical analyses using R, most of the mathematics we come across in the book isn't done for its own sake, but is rather required for implementing the given statistical procedure.

R can manage operations on sets quite easily as well. For example, we can define two sets of elements as follows:

```
> set.A <- c(1, 2, 3, 4, 5)
> set.B <- c(2, 3, 4, 5, 6)
```

The **union** of these two sets, which is defined as elements in set A or set B or both, can be computed by:

```
> union(set.A, set.B)
[1] 1 2 3 4 5 6
```

The **intersection** of these two sets, defined as elements in set A and in set B, is computed by:

```
> intersect(set.A, set.B)
[1] 2 3 4 5
```

Notice that the intersection only contains elements that are in both sets A and B.

We may be interested in knowing whether the elements of one set are contained within the other element-by-element:

```
> set.A %in% set.B
[1] FALSE  TRUE  TRUE  TRUE  TRUE
```

The first element is FALSE because the number 1 in set A does not occur in set B. All other values are TRUE because the remaining numbers in set A occur in set B. Two sets are **equal** if they have the same elements. We can test for set equality using the setequal() function (left as an exercise).

2.2.2 Plotting Curves Using curve()

While we are on the topic of mathematics, it behooves us to briefly survey the kinds of functions R can plot. The curve() function is especially useful for this purpose. Here's how it works: one first enters a **polynomial** defined by the variable x, then specifies the range of values that we would like the domain to take on. The **domain** of the function is the set of possible values that can be input into the function. For example, consider the following polynomial:

$$f(x) = 4x^2 + 3$$

This is a **quadratic function**, which we can easily write in R, 4*x^2 + 3. Now, let's specify the range of values for the domain of the function. Let's make the function take on only positive values from 0 to 100. So, once we add the `curve()` function to this, we have:

```
> curve(4*x^2+3, 0, 100)
```

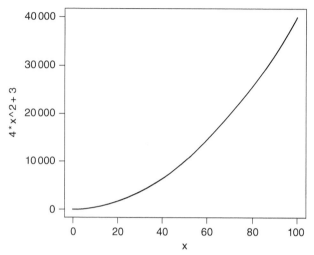

At first glance, the curve doesn't look much like a parabola, but that's due to how we specified the range. If we shrink the range down to −1 to 1, we get:

```
> curve(4*x^2+3, -1, 1)
```

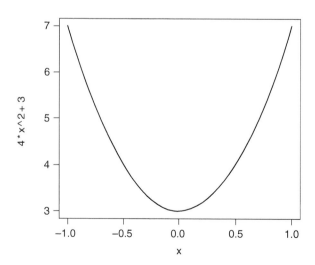

The curve now resembles the parabola with which we are more familiar.

Polynomials, as we will see later in the book, can be used to fit data in regression. If a **line** is the best fit, then the method is that of **linear regression**. Recall that a line is simply a polynomial of degree 1. If a different polynomial is theorized to fit the data better, then **polynomial regression** becomes an option, such as that featured below in which we see that a polynomial of degree 3 ("cubic polynomial") is a reasonable fit to the data:

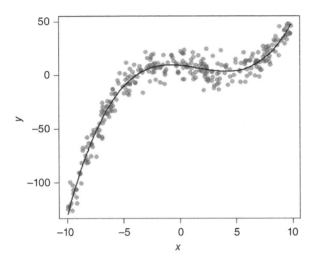

2.3 Vectors and Matrices in R

A basic understanding of vectors and matrices is essential for understanding modern statistics and data analysis. Thankfully, most of the concepts are very intuitive and can be explained using demonstrations in R.

A **vector** can be defined as either a **list of numbers**, which is the typical computer science definition of the term, or as a **line segment having magnitude and direction**, which is the typical physics definition of the word. When we usually define a vector in the course of data analysis, we are simply aggregating a series of observations on some variable. Technically, a vector can also consist of only a single number, in which case it is usually referred to as a **scalar**. There are a few different ways of creating a vector in R. The most common way is to simply list the name you wish to give to the vector, followed by the array of numbers assigned to that vector. For example, the following assigns the name x to the list of numbers 1, 3, 5, 7, 9:

```
> x <- c(1, 3, 5, 7, 9)
> x
[1] 1 3 5 7 9
```

We can verify that it is a vector (Wickham, 2014) by:

```
> x <- c(1, 3, 5, 7, 9)
> is.atomic(x) || is.list(x)
[1] TRUE
```

We could have also, had we really wanted to, named the vector at the end rather than at the beginning:

```
> c(1, 3, 5, 7, 9) -> x
> x
[1] 1 3 5 7 9
```

Or, we could have made it more explicit that we are "assigning" the name x to the list of numbers:

```
> assign("x", c(1, 3, 5, 7, 9))
> x
[1] 1 3 5 7 9
```

Whichever way we choose, we note that the vector x has 5 elements. We can use the length() function in R to tell us so:

```
> length(x)
[1] 5
```

Suppose we wanted to add a new element to our vector in a particular position in the list, such as adding an 8 between numbers 7 and 9. We could accomplish this using the following:

```
> x <- c(x[1:4], 8, x[5])
> x
[1] 1 3 5 7 8 9
```

We see that the above added the number 8 in the fifth position in the vector, as indicated by x[5]. We can also extract elements of a vector by **indexing**. For example, for vector x, suppose we wish to extract elements 2 and 4, which correspond to numbers 3 and 7 in the vector:

```
> x[c(2, 4)]
[1] 3 7
```

We can also exclude elements of the vector. Suppose we wish to exclude the last element of x, which corresponds to the number 9. Notice that 9 is in the sixth position, so we wish to remove the sixth element:

```
> x[-6]
[1] 1 3 5 7 8
```

If we wanted to remove elements 2 to 4 inclusive, we would code:

```
> x[-2:-4]
[1] 1 8 9
```

Let's define another vector, y:

```
> y <- c(2, 4, 6, 8, 10)
> y
[1]  2  4  6  8 10
```

The vector y also has length 5:

```
> length(y)
[1] 5
```

The addition of vectors means adding the first element of one vector to the first element of the other vector, and likewise the second element of one vector to the second element of the other vector, etc. That is, vector addition takes place **elementwise** across the two (or more) vectors (using our original vector x):

```
> x + y
[1]  3  7 11 15 19
```

In linear algebra, only vectors with the same number of elements can be added. If we try to add vectors of different lengths in R, it will begin to **recycle** the elements of the shorter vector. Hence, if vector y had elements 2, 4, 6, 8, 10, 12, and we tried adding y to x, we would get:

```
> x <- c(1, 3, 5, 7, 9)
> y <- c(2, 4, 6, 8, 10, 12)
> x + y
[1]  3  7 11 15 19 13
Warning message:
In x + y : longer object length is not a multiple of
shorter object length
```

As we can see, R correctly added the elements of the first 5 positions in each vector (i.e. $1 + 2 = 3$, $3 + 4 = 7$, etc.). But then when it got to element 6 in y, equal to 12, it had no corresponding sixth element in x to add it to. So it began recycling elements, and added 12 to the first element of x, to get 13.

We can also compute a **cumulative sum** of a vector using `cumsum()`. For example, the cumulative sum of vector x is computed:

```
> x <- c(1, 3, 5, 7, 9)
> cumsum(x)
[1]   1   4   9 16 25
```

Notice that the sum of the first two elements is 4, then the first three elements 9, and so on. The **cumulative product** can also be computed using `cumprod()`.

When we speak of the **dot product** between two vectors, we are referring to the **sum of each element multiplied pairwise**. For example, for our vectors x and y we define the dot product as:

```
> x <- c(1, 3, 5, 7, 9)
> y <- c(2, 4, 6, 8, 10)
> x%*%y
     [,1]
[1,]  190
```

To demonstrate how this was computed, we compute it the long way:

```
> dot.product <- 1*2 + 3*4 + 5*6 + 7*8 + 9*10
> dot.product
[1] 190
```

Notice that to get the sum, we added respective products of each element of the first vector with corresponding elements of the second vector. We could have also computed the dot product using `crossprod(x, y)`.

Sometimes we want to scale a vector by, not surprisingly, a **scalar**! That is, we multiply the vector element-by-element by a particular number. Let's scale vector x by a factor of 2:

```
> 2*x
[1]   2   6 10 14 18
```

Notice that each element of the vector was multiplied by the scalar 2. This has the effect of elongating the vector, since each element of it has increased by a factor of 2.

2.4 Matrices in R

When we speak of a **matrix**, we are referring to an array of vectors contained in the same object. As we will see later, an understanding of matrices is required for making sense of relatively advanced statistical models in multivariable and multivariate settings. We will revisit matrices as we need them and as we proceed through the book. For now, we wish to simply demonstrate how we can construct a matrix, and compute a few elementary matrix operations on it.

We can create a matrix with numbers 1 through 8, having 4 rows and 2 columns:

```
> S <- matrix(1:8, 4, 2)
> S
      [,1] [,2]
[1,]    1    5
[2,]    2    6
[3,]    3    7
[4,]    4    8
```

where S is the name of our matrix, 1:8 designates the matrix to have entries 1–8, that is, to have numbers in the matrix ranging from 1 to 8, and 4,2 requests the matrix to have 4 rows and 2 columns. If we specify too many elements for our matrix but too few columns and/or rows, R will simply **truncate** the matrix. For example, the following asks for a matrix with values 20 through 29, but only specifies a 2 × 2 matrix. Consequently, R simply inputs the first 4 values of the matrix and ignores the rest of the numbers:

```
> S <- matrix(20:29, 2, 2)
> S
      [,1] [,2]
[1,]   20   22
[2,]   21   23
```

The **dimension** of a matrix is the number of rows by the number of columns. For our matrix S just above, we have 2 rows and 2 columns, so the dimension is equal to 2 2:

```
> dim(S)
[1] 2 2
```

Just as we did for vectors, we can also index matrices. For example, suppose for our matrix S we wished to extract the second column:

```
> S[,2]
[1] 22 23
```

In statistical procedures such as **principal component analysis** and **factor analysis**, we will be required to enter a covariance or correlation matrix as an input to the analysis, and hence will often have to build such a matrix so that we can run the statistical method. But what is a covariance matrix? A **covariance matrix**, sometimes called **variance–covariance matrix**, is a matrix of several variables with variances along the main diagonal of the matrix (i.e. from top left to bottom right), and covariances along the off-diagonal, where "E" is the expectation operator:

$$
\mathbf{K}_{XX} =
\begin{bmatrix}
E[(X_1 - E[X_1])(X_1 - E[X_1])] & E[(X_1 - E[X_1])(X_2 - E[X_2])] & \cdots & E[(X_1 - E[X_1])(X_1 - E[X_n])] \\
E[(X_2 - E[X_2])(X_1 - E[X_1])] & E[(X_2 - E[X_2])(X_2 - E[X_2])] & \cdots & E[(X_2 - E[X_2])(X_n - E[X_n])] \\
\vdots & \vdots & \ddots & \vdots \\
E[(X_n - E[X_n])(X_1 - E[X_1])] & E[(X_n - E[X_n])(X_2 - E[X_2])] & \cdots & E[(X_n - E[X_n])(X_n - E[X_n])]
\end{bmatrix}
$$

The **correlation matrix** is a standardized covariance matrix, in which respective covariances are divided by the product of standard deviations (see Chapter 4 for a brief discussion of covariance and correlation):

$$
\mathrm{corr}(\mathbf{X}) =
\begin{bmatrix}
1 & \dfrac{E[(X_1 - \mu_1)(X_2 - \mu_2)]}{\sigma(X_1)\sigma(X_2)} & \cdots & \dfrac{E[(X_1 - \mu_1)(X_n - \mu_n)]}{\sigma(X_1)\sigma(X_n)} \\
\dfrac{E[(X_2 - \mu_2)(X_1 - \mu_1)]}{\sigma(X_2)\sigma(X_1)} & 1 & \cdots & \dfrac{E[(X_2 - \mu_2)(X_n - \mu_n)]}{\sigma(X_2)\sigma(X_n)} \\
\vdots & \vdots & \ddots & \vdots \\
\dfrac{E[(X_n - \mu_n)(X_1 - \mu_1)]}{\sigma(X_n)\sigma(X_1)} & \dfrac{E[(X_n - \mu_n)(X_2 - \mu_2)]}{\sigma(X_n)\sigma(X_2)} & \cdots & 1
\end{bmatrix}
$$

We can build the covariance or correlation matrix in R in a few ways. We can read it off an external file, or, for a matrix of relatively small dimension, simply code in the vectors then join them together. The following is an example of a correlation matrix, one we will use again later in the book. We first define each row by "concatenating" elements using c(). Then, we name the object cormatrix, and bind the columns of the matrix together using cbind():

```
> c1 <- c(1.000, 0.343, 0.505, 0.308, 0.693, 0.208, 0.400, 0.455)
> c2 <- c(0.343, 1.000, 0.203, 0.400, 0.187, 0.108, 0.386, 0.385)
> c3 <- c(0.505, 0.203, 1.000, 0.398, 0.303, 0.277, 0.286, 0.167)
> c4 <- c(0.308, 0.400, 0.398, 1.000, 0.205, 0.487, 0.385, 0.465)
> c5 <- c(0.693, 0.187, 0.303, 0.205, 1.000, 0.200, 0.311, 0.485)
> c6 <- c(0.208, 0.108, 0.277, 0.487, 0.200, 1.000, 0.432, 0.310)
> c7 <- c(0.400, 0.386, 0.286, 0.385, 0.311, 0.432, 1.000, 0.365)
> c8 <- c(0.455, 0.385, 0.167, 0.465, 0.485, 0.310, 0.365, 1.000)
> cormatrix <- cbind(c1, c2, c3, c4, c5, c6, c7, c8)

> cormatrix
        c1    c2    c3    c4    c5    c6    c7    c8
[1,] 1.000 0.343 0.505 0.308 0.693 0.208 0.400 0.455
[2,] 0.343 1.000 0.203 0.400 0.187 0.108 0.386 0.385
[3,] 0.505 0.203 1.000 0.398 0.303 0.277 0.286 0.167
[4,] 0.308 0.400 0.398 1.000 0.205 0.487 0.385 0.465
[5,] 0.693 0.187 0.303 0.205 1.000 0.200 0.311 0.485
```

```
[6,]  0.208 0.108 0.277 0.487 0.200 1.000 0.432 0.310
[7,]  0.400 0.386 0.286 0.385 0.311 0.432 1.000 0.365
[8,]  0.455 0.385 0.167 0.465 0.485 0.310 0.365 1.000
```

Let's now compute a few functions on our newly created correlation matrix. Notice first that the main diagonal is made up of values of 1, since for each pairing, the correlation of each variable with itself is equal to 1. The sum of values along the main diagonal of the matrix is referred to as the **trace** of the matrix, which we can compute as `tr()` in R. We first load the package **psych**:

```
> library(psych)
> tr(cormatrix)
[1]  8
```

The trace is equal to 8, since there are 8 variables and hence 8 values of 1 along the main diagonal, which when we sum them, equals a total of 8. The dimension of the matrix should equal 8 by 8, as we can readily verify:

```
> dim(cormatrix)
[1]  8 8
```

We can confirm that there are only values of 1 along the main diagonal using the `diag()` function, which will reproduce the values on that main diagonal:

```
> diag(cormatrix)
[1] 1 1 1 1 1 1 1 1
```

2.4.1 The Inverse of a Matrix

Recall that in ordinary scalar algebra, the inverse of a scalar a is defined as a^{-1} so that the following relation holds true for any value of $a \neq 0$:

$$a \cdot a^{-1} = \frac{a}{1} \cdot \frac{1}{a} = \frac{a}{a} = 1$$

In matrix algebra, the inverse is analogously defined, but in terms of matrices: $AB = BA = I$. We say B is the inverse of matrix A. In R, we can compute the inverse of a matrix using the function `solve()`. For example, the inverse of `cormatrix` is computed as:

```
> I <- solve(cormatrix)
> I
           [,1]        [,2]        [,3]        [,4]        [,5]        [,6]
c1   2.62391779 -0.32529402 -0.79748547  0.07143344 -1.34883744  0.14277980
c2  -0.32529402  1.45211520  0.03592592 -0.40955548  0.22539180  0.31041745
c3  -0.79748547  0.03592592  1.56495953 -0.48016535  0.04087805 -0.13719304
c4   0.07143344 -0.40955548 -0.48016535  1.85491453  0.19261836 -0.58606745
c5  -1.34883744  0.22539180  0.04087805  0.19261836  2.14973569 -0.07659356
```

```
c6   0.14277980   0.31041745  -0.13719304  -0.58606745  -0.07659356   1.53618210
c7  -0.27074036  -0.38009280  -0.03786275  -0.05349807  -0.06236400  -0.49852528
c8  -0.25993622  -0.29342246   0.34747695  -0.54938320  -0.56556392  -0.14614977
              [,7]           [,8]
c1  -0.27074036  -0.25993622
c2  -0.38009280  -0.29342246
c3  -0.03786275   0.34747695
c4  -0.05349807  -0.54938320
c5  -0.06236400  -0.56556392
c6  -0.49852528  -0.14614977
c7   1.55055676  -0.08044157
c8  -0.08044157   1.77763927
```

As we will see later in the book, if the inverse is not computable, then it often causes computational difficulties in carrying out statistical procedures.

The **determinant** of a matrix is a number associated with every square matrix that is unique to that matrix. Though determinants can be computed for any square dimension matrix, the easiest case for demonstration is that for a 2×2 matrix. A determinant of a 2×2 matrix is defined as follows:

$$|\mathbf{A}| = \begin{vmatrix} a_{11} & a_{12} \\ a_{21} & a_{22} \end{vmatrix} = a_{11}a_{22} - a_{12}a_{21}$$

We can compute the determinant in R quite easily, using the det () function. For cormatrix, for example, the determinant is:

```
> det(cormatrix)
[1] 0.06620581
```

In terms of applied statistical procedures, determinants equal to 0 will typically cause difficulty and in the world of matrix and linear algebra, has a whole bunch of conditions and consequences associated along with it. In brief, determinants equal to 0 are "bad news" and will typically generate error messages in common statistical procedures such as regression and others, because it usually implies a key matrix cannot be inverted (which causes a lot of problems for the mathematics underlying the statistical method).

A square matrix is regarded as **orthogonal** if the following condition holds: $\mathbf{AA}' = \mathbf{A}'\mathbf{A} = \mathbf{I}$, where $\mathbf{I}$ is the **identity matrix**, having values of 1 along the main diagonal and zeroes everywhere else. For instance, an identity matrix of 3×3 dimension is the following:

$$\begin{bmatrix} 1 & 0 & 0 \\ 0 & 1 & 0 \\ 0 & 0 & 1 \end{bmatrix}$$

The identity matrix in linear algebra is analogous to the unit scalar of 1 in ordinary scalar algebra. As we will see in factor analysis, we often wish to **rotate a matrix** using rotational methods such as **varimax rotation**.

2.4.2 Eigenvalues and Eigenvectors

Many statistical analyses, in one way or another, come down to solving for eigenvalues and eigenvectors of a correlation or covariance matrix. Now, this is more obvious in some procedures such as principal component analysis, but it is also true that eigenvalues and eigenvectors "show up" in many other statistical methods such as regression, discriminant analysis, and MANOVA. And while an in-depth study of them is beyond the scope of this book (and most others on applied or computational statistics), a brief overview of them is useful to get the main ideas across.

For every square matrix $\mathbf{A}$, it is a mathematical fact that a scalar λ and vector $\mathbf{x}$ can be obtained so that the following equality holds true:

$$\mathbf{Ax} = \lambda\mathbf{x}$$

But what does the above equality say "in English?" It says that if we perform a **transformation** on the vector $\mathbf{x}$ via the matrix $\mathbf{A}$, this is equivalent (i.e. equal sign) to transforming the vector $\mathbf{x}$ by multiplying it by a "special scalar" λ. This very special scalar is called an **eigenvalue** of the matrix $\mathbf{A}$, and $\mathbf{x}$ is called an **eigenvector**. That is, the transformation on the eigenvector is equivalent to multiplying that vector by the scalar λ. The ramifications and generality of this equation are quite deep at a technical level, and eigenvalues and eigenvectors play a significant role in many branches of mathematics. They also correspond to many physical concepts in many scientific fields such as physics. Since applied statistics, especially multivariate statistics, largely has matrix and linear algebra as its foundations, **eigen analysis**, the process of obtaining the eigenvalues and eigenvectors from a correlation or covariance matrix, is central to many of these statistical procedures.

For example, in a principal component analysis, eigenvalues will refer to the actual **variances of components**, and elements of the eigenvectors will refer to the weights from which the respective linear combinations (i.e. "components") are formed. Since this is primarily an applied book, we will revisit these concepts only later in the book as we need them, and tell you exactly what you need to know to interpret them within the given statistical method.

For now, it is still worth demonstrating how we can perform an eigen analysis on a matrix. We will use `cormatrix` to demonstrate the extraction of eigenvalues and eigenvectors:

```
> eigen(cormatrix)
eigen() decomposition
$`values`
[1] 3.4470520 1.1572358 0.9436513 0.8189869 0.6580753 0.3898612 0.3360577
[8] 0.2490798

$vectors
            [,1]         [,2]          [,3]        [,4]         [,5]          [,6]
[1,] -0.4125682 -0.45771854 -0.098873465  0.08795002  0.066755981 -0.1352881
[2,] -0.3034726  0.11443307  0.637320040  0.47138991  0.088175982 -0.4697628
[3,] -0.3180940 -0.06869755 -0.546391282  0.58136533 -0.121757436  0.1921707
[4,] -0.3730602  0.43006317 -0.001725853  0.11149001 -0.471416291  0.1757057
[5,] -0.3572744 -0.54341592 -0.067983885 -0.31379342  0.005351703 -0.2192800
[6,] -0.3008318  0.49347718 -0.380799221 -0.40057953  0.065440460 -0.5409419
[7,] -0.3664721  0.21605328  0.060830978 -0.06123129  0.775010839  0.4437028
[8,] -0.3806410 -0.04717545  0.363552650 -0.39618322 -0.381782069  0.3945174
            [,7]         [,8]
[1,]  0.0911831  0.75610613
[2,] -0.1315217 -0.14378501
[3,] -0.3684690 -0.26465820
[4,]  0.6393505  0.03954700
[5,]  0.3112839 -0.57354688
[6,] -0.2453502  0.05840491
[7,]  0.1046301 -0.05574971
[8,] -0.5116712  0.02337248
```

The eigenvalues of the matrix are thus 3.447, 1.157, 0.943, 0.819, 0.658, 0.389, 0.336, and 0.249. Notice they are reported in decreasing values of their magnitude. When we perform a principal component analysis on this matrix, we will see that while there are other facets to components analysis we will need to attend to, we are, in fact, performing an eigen analysis on a correlation or covariance matrix, and will revisit these exact eigenvalues when we run the procedure.

The $vectors section of the output are the eigenvectors, each associated with its respective eigenvalue. For example, the first eigenvector is that of

```
-0.4125682
-0.3034726
-0.3180940
-0.3730602
-0.3572744
-0.3008318
-0.3664721
-0.3806410
```

Can we demonstrate to ourselves that this first eigenvector must correspond to the first eigenvalue? Absolutely. Recall our earlier equation, that of $\mathbf{Ax} = \lambda\mathbf{x}$. If R computed the eigenvalues and eigenvectors correctly, then this equation should hold. Let's try computing it manually, by multiplying the eigenvector by the correlation matrix to get the left-hand side of the equation, then multiplying the eigenvector by the eigenvalue to get the right-hand side. Let's first build up the eigenvector:

```
> eigen.vector <- c(-0.4125682, -0.3034726, -0.3180940,
-0.3730602, -0.3572744, -0.3008318, -0.3664721, -0.3806410)
```

Now, let's multiply this vector by cormatrix, which corresponds to matrix **A**:

```
> Ax = cormatrix%*%eigen.vector
> Ax
             [,1]
[1,]  -1.422144
[2,]  -1.046086
[3,]  -1.096486
[4,]  -1.285958
[5,]  -1.231543
[6,]  -1.036983
[7,]  -1.263249
[8,]  -1.312089
```

Now, we can easily demonstrate that this equals the right-hand side of $\lambda\mathbf{x}$:

```
> lambda.x = 3.4470520*eigen.vector
> lambda.x
[1]  -1.422144 -1.046086 -1.096487 -1.285958 -1.231543
-1.036983 -1.263248
[8]  -1.312089
```

We can confirm that we obtained the same vector, and hence the demonstration is complete. We should take this opportunity to note that we used the word "demonstration" here on purpose, in that it is not a justification or "proof" by any means, but rather simply a particular showing that the equation works with a few numbers. **Proofs in mathematics** require much more than simply demonstrating that things happen to work with a few numbers. They generally require one to justify that the relation holds universally for a given set of numbers (e.g. all real numbers).

2.5 How to Get Data into R

There are several ways of getting data into R. For small data sets, often of the type featured in this book for demonstrations of statistical techniques, one can enter data "on the spot" through generating vectors, then joining these vectors together to form a **data frame**. As an example of a data set that we will use in this book, consider hypothetical data on quantitative ability, verbal ability, and a training group designating whether individuals received no training (1), some (2), or much training (3) in learning these skills (Table 2.1).

In a moment, we will ask R to load this data from our local hard drive into R's work space, but for now, imagine we did not have the data externally saved. How would we enter it? Since it's a very small data set, we can simply build the columns for each variable, then generate the data frame. Each column of the data set is a list or vector. To generate the quant vector for instance, we compute:

```
> quant <- c(5, 2, 6, 9, 8, 7, 9, 10, 10)
> quant
[1]   5   2   6   9   8   7   9  10  10
```

Again, notice the format for generating the vector. We start off with naming the object we wish to create, that of quant, then list the data points in parentheses, but with "c" before it, so that we are joining or **concatenating**, as R likes

Table 2.1 Hypothetical data on quantitative and verbal ability as a function of training.

Subject	Quantitative	Verbal	Training
1	5	2	1
2	2	1	1
3	6	3	1
4	9	7	2
5	8	9	2
6	7	8	2
7	9	8	3
8	10	10	3
9	10	9	3

1 = no training, 2 = some training, 3 = extensive training.

to call it, the list. To verify we produced it correctly, we simply make a call to quant and it generates the vector for us. We can build verbal and train accordingly:

```
> verbal <- c(2, 1, 3, 7, 9, 8, 8, 10, 9)
> train <- c(1, 1, 1, 2, 2, 2, 3, 3, 3)
```

Now that we've produced the vectors, we are ready to join them into a data frame. We can do this one of two ways. One way is to use cbind() ("column bind") to join the columns:

```
> iq.train <- cbind(quant, verbal, train)
> iq.train
      quant verbal train
[1,]      5      2     1
[2,]      2      1     1
[3,]      6      3     1
[4,]      9      7     2
[5,]      8      9     2
[6,]      7      8     2
[7,]      9      8     3
[8,]     10     10     3
[9,]     10      9     3
```

Or, we could have also generated the data frame by reference to the function data.frame(), which is usually the preferred way of generating data sets in R:

```
> iq.train <- data.frame(quant, verbal, train)
> iq.train
  quant verbal train
1     5      2     1
2     2      1     1
3     6      3     1
4     9      7     2
5     8      9     2
6     7      8     2
7     9      8     3
8    10     10     3
9    10      9     3
```

Having entered the data, we are now ready to perform analyses on the object iq.train.

Now, as you might imagine, entering data this way works well for very small data sets. Most often of course, data sets will be much larger and stored in external files. We of course need a way to access that data and bring it into R. A common way of doing this is through the `read.table()` function, which locates the data where you have it stored on your computer, and loads it into R. For the following demonstration, we have the data file stored as a **text file (.txt)** generated using **notepad** or another **text editor**. This would be the file you constructed manually external to R, or that you were provided for data analysis from another party. We will ask R to load this data, so it is accessible for analysis.

Before we do so, we need to tell R where to locate the file. The "Graphical User Interface" (GUI) for this is the easier way to get this done. Go to the top corner of the R window to **File**, then scroll down to **Change dir**, and select the path where your data are found. For instance, if the text file is on your Desktop, then you'll most likely want to access disk **C:** (most Desktops on computers are located on disk **C:**) then follow the path to your Desktop. Once you have located the Desktop, R will be able to search it to locate the file. You can then ask R to read the data file through `read.table()`. For convenience, we have named the text file **iqtrain.txt**. The following code will bring in our text file:

```
> iq.train <- read.table("iqtrain.txt", header = T)
> iq.train
  quant verbal train
1     5      2     1
2     2      1     1
3     6      3     1
4     9      7     2
5     8      9     2
6     7      8     2
7     9      8     3
8    10     10     3
9    10      9     3
```

A couple more things to note about reading in the data file using `read.table()`. Be sure when identifying the data set, you have " " around it, and the comma is just on the **outside** of the quotation marks and not on the inside. The `header = T` statement tells R that there are variable name headers in the data, and to read the first row of the data as the column names, and not actual data points. If we had no variable names in our data, we could have either left the statement out, or coded `header = F`. Without variable name headers, R would have just used generic "V" labels to identify variables. To read a data file in comma-separated values (CSV), we can use `read.csv()`.

2.6 Merging Data Frames

Sometimes we wish to merge two data frames into a single one. As a trivial example, suppose we wished to merge the data frame **iris** with that of **GaltonFamilies**:

```
> iris.GaltonFamilies <- merge(iris, GaltonFamilies)
> iris.GaltonFamilies
> head(iris.GaltonFamilies)
```

	Sepal.Length	Sepal.Width	Petal.Length	Petal.Width	Species	family	father
1	5.1	3.5	1.4	0.2	setosa	001	78.5
2	4.9	3.0	1.4	0.2	setosa	001	78.5
3	4.7	3.2	1.3	0.2	setosa	001	78.5
4	4.6	3.1	1.5	0.2	setosa	001	78.5
5	5.0	3.6	1.4	0.2	setosa	001	78.5
6	5.4	3.9	1.7	0.4	setosa	001	78.5

	mother	midparentHeight	children	childNum	gender	childHeight
1	67	75.43	4	1	male	73.2
2	67	75.43	4	1	male	73.2
3	67	75.43	4	1	male	73.2
4	67	75.43	4	1	male	73.2
5	67	75.43	4	1	male	73.2
6	67	75.43	4	1	male	73.2

We see that R combined both data frames into a single one. We requested `head()` to get the first few cases. It is also possible to access data from online sources via URLs. For an example of this, see Matloff (2011), who also provides a good discussion of implementing **parallel R**, which is appropriate for situations in which memories of several computers are pooled to facilitate some statistical analyses having exceedingly large runtimes.

2.7 How to Install a Package in R, and How to Use It

One thing that makes R a truly great software is not only that it is free, but also that several thousand people are continually, day and night, working to improve the functionality and capability of the software. Veritable R experts in the truest sense of the word immensely enjoy computational statistics, and daily create what are known as **packages** that allow the user to do things in the software that would be difficult or impossible to do in more conventional software such as SPSS or SAS. These packages can be loaded into R as needed to either facilitate the use of the software for basic statistical tasks, or to conduct relatively specialized functions for advanced analyses. For instance, later in the book when we survey ANOVA models, we will require the lme4 package to run random effects and mixed models. This package is not available automatically with R, that is, it is not available with the **base version** of R. The user has to install it individually and separately in order to be able to use it. Another package that has many useful functions is the car package. To install this package, we simply code the following:

```
> install.packages("car")
```

Upon making this request, R will then ask you to choose a **mirror site** from which to download the package. Once more, feel free to choose a mirror site relatively close to your geographical location, but usually any mirror site will do. If you wish to download more than a single package at a time, you could also **concatenate** them into a list as follows, where we load both the **car** and **MASS** packages at once:

```
> install.packages(c("car", "MASS"))
```

Once you've selected the relevant package, it will download into R, and you will be able to use it. Depending on the complexity of the package, downloading it could take a bit of time, whereas simpler packages download much more quickly. To bring the package into your search path, you need the library() function:

```
> library(car)
>
```

The package is now available for use. Each package comes with R documentation that can be found online that details the package's functionality. Be forewarned, these documents are sometimes heavy reading and can span many pages, but if you truly start to use a package extensively, you will want to familiarize yourself with its documentation. Usually, researchers won't study the whole package; they will simply scroll down to the part they need, and see if they can make it work in R. For instance, if we wanted to access the car package, we can simply **Google**, and look for the **PDF** file:

Google   Car package R   [PDF] Package 'car' - The R Project for Statistical Computing
https://cran.r-project.org/web/packages/car/car.pdf ▾
by J Fox · 2018 · Cited by 132 · Related articles
Aug 23, 2018 · **Package** 'car'. August 24, 2018. Version 3.0-2. Date 2018-08-23. Title Companion to Applied Regression. Depends R (>= 3.2.0), carData ...

You can then open up the PDF file, and for this package, you will see it is a total (as of this writing) of 147 pages! The document will have some descriptive details about the package, and then you're off trying to figure out how to use it. Remember once more, working with code and figuring out a computing language is very much trial-and-error; you will likely experience much of this when trying to implement a new package – do not despair, the feeling of "getting nowhere" is normal from time to time. It usually takes longer to get a package working properly than anticipated, especially for the more complicated ones. We can also request more help on a particular package:

```
> library(car)
> library(help = car)
```

R Documentation for package 'car'

scatter3d	Three-Dimensional Scatterplots and Point Identification
scatterplot	Enhanced Scatterplots with Marginal Boxplots, Point Marking, Smoothers, and More
scatterplotMatrix	Scatterplot Matrices
showLabels	Functions to Identify and Mark Extreme Points in a 2D Plot.
sigmaHat	Return the scale estimate for a regression model
some	Sample a Few Elements of an Object
spreadLevelPlot	Spread-Level Plots
subsets	Plot Output from regsubsets Function in leaps package
symbox	Boxplots for transformations to symmetry
testTransform	Likelihood-Ratio Tests for Univariate or Multivariate Power Transformations to Normality
vif	Variance Inflation Factors
wcrossprod	Weighted Matrix Crossproduct
whichNames	Position of Row Names

```
Further information is available in the following vignettes in
directory 'C:/Users/Dan Denis/Documents/R/win-library/3.5/car/doc':

embedding: Using car functions inside user functions (source, pdf)
```

Before you think you're getting over your head after looking at the documentation for the **car** package and others, keep in mind that for most of this book and for many of your analyses, though you may need some of the functions in a given package, that's what this book is for and books like this one. They tell you exactly what you need from the given package so you don't have to sift through the documentation and figure it out for yourself. So, as you work through the book and come across an area where I tell you you'll need a certain package, all you need to do is follow my guidance and I will tell you what you need to code from the given package. Nobody, not even very experienced R users, know all the packages or all their functions, but if you go on to use R more extensively, you can expect to work on implementing several of them. As of this writing, there are currently approximately 10,000 packages available for download, and that number continues to grow everyday. Again, this is one advantage of having open-source software that is available "live" on the internet. Contributions to it can occur **moment to moment**, rather than **version to version** such as with traditional software.

If you would like to know which packages are currently loaded in your workspace, the search() function will display the list, as well as other objects currently loaded in your workspace:

```
> search()
[1] ".GlobalEnv"          "Galton"
[3] "faithful"            "package:pwr"
[5] "learn"               "package:nlme"
[7] "package:lme4"        "package:Matrix"
[9] "package:phia"        "package:PMCMR"
```

If you run search() on your machine, you will not see the same packages as above, as I have already loaded these specifically for a few of the tasks and functions I specifically needed. However, if you did install the car package successfully, it should show up on the above list.

Packages in R give the software extended capabilities and offer several specialized functions for conducting very specific tasks and statistical computations not available in the base package for R. As you get to know the software better, and conduct increasingly complex statistical analyses, you will likely be downloading several packages as you work with R. The availability and implementation of packages in R is part of what makes the software truly remarkable.

2.8 How to View the Top, Bottom, and "Some" of a Data File

Especially when we load a very large file into R, much too large to see the whole thing at once, we would like a way to visualize the top, bottom, and maybe a few random observations within the data frame to get a feel for it. As an example, let's load the iris data into R, a data frame we will use throughout the book. The iris data is part of the base package in R, and is available by simply typing it at the command prompt:

```
> iris
```

Now when you type this, you'll get 150 observations. We will describe the data later in the book as we use it. For now, we are only interested in demonstrating how we can view parts of it. To see only the top portion of the data frame, we can use the head() function:

```
> head(iris)
   Sepal.Length Sepal.Width Petal.Length Petal.Width Species
1           5.1         3.5          1.4         0.2  setosa
2           4.9         3.0          1.4         0.2  setosa
3           4.7         3.2          1.3         0.2  setosa
4           4.6         3.1          1.5         0.2  setosa
5           5.0         3.6          1.4         0.2  setosa
6           5.4         3.9          1.7         0.4  setosa
```

The head() function produced the first 6 observations of the data (listed 1 through 6 above). We can likewise get a glimpse of the last few data points in the dataframe using tail(), which will show us the last 6 observations in the data:

```
> tail(iris)
    Sepal.Length Sepal.Width Petal.Length Petal.Width   Species
145          6.7         3.3          5.7         2.5 virginica
146          6.7         3.0          5.2         2.3 virginica
147          6.3         2.5          5.0         1.9 virginica
148          6.5         3.0          5.2         2.0 virginica
149          6.2         3.4          5.4         2.3 virginica
150          5.9         3.0          5.1         1.8 virginica
```

Finally, using the **car** package that we just installed, we can get a glimpse of some () of the data. Notice that since the some () function comes from the car package, we first make sure the package is available by first coding library(car). Now, if you know car is already in R's search path, then you don't have to call it every time, but if you're not sure, or you just installed it and are using it for the first time, then library(car) is necessary, otherwise some(iris) will generate an error:

```
> library(car)
> some(iris)
    Sepal.Length Sepal.Width Petal.Length Petal.Width    Species
3            4.7         3.2          1.3         0.2     setosa
4            4.6         3.1          1.5         0.2     setosa
20           5.1         3.8          1.5         0.3     setosa
30           4.7         3.2          1.6         0.2     setosa
68           5.8         2.7          4.1         1.0 versicolor
105          6.5         3.0          5.8         2.2  virginica
115          5.8         2.8          5.1         2.4  virginica
```

```
131           7.4          2.8          6.1          1.9    virginica
145           6.7          3.3          5.7          2.5    virginica
147           6.3          2.5          5.0          1.9    virginica
```

As we can see, the function `some(iris)` generated for us a random sample of 10 observations from the data set, spanning a minimum case number of 3 up to a maximum case of 147, and random cases in between.

2.9   How to Select Subsets from a Dataframe

Suppose we would like to select the variable `quant` and `verbal` from the `iq.train` data. We can first ask R to tell us what variables are available in the given data frame:

```
> names(iq.train)
[1] "quant"  "verbal" "train"
```

Let's now extract variables `quant` and `verbal`:

```
> iq.train[1:2]
   quant verbal
1      5       2
2      2       1
3      6       3
4      9       7
5      8       9
6      7       8
7      9       8
8     10      10
9     10       9
```

The [1:2] communicates to R to pull out variables in columns 1 and 2. If we only wanted to work with these two variables, it may be worth defining a new object that contains only these two:

```
> vq <- iq.train[1:2]
> vq
   quant verbal
1      5       2
2      2       1
3      6       3
4      9       7
5      8       9
```

6	7	8
7	9	8
8	10	10
9	10	9

We could have also accomplished the above using the `select()` function:

```
> library(dplyr)
> select(iq.train, quant, verbal)
```

To call a specific variable in a data frame, we use the $ symbol. For instance, suppose we wanted the `quant` variable only, we could code:

```
> iq.train$quant
[1]   5   2   6   9   8   7   9 10 10
```

Maybe we want to select only those cases having `quant` scores greater than 8:

```
> iq.train[iq.train$quant > 8,]
    quant verbal train
4      9      7     2
7      9      8     3
8     10     10     3
9     10      9     3
```

In the above code, notice we are first identifying the data frame within brackets, then specifying the variable within the data frame through $quant, before identifying that we want cases with values greater than 8. We can also use the `filter()` function in the **dplyr** package to select cases or groups of cases:

```
> library(dplyr)
> filter(iq.train, quant == 10, verbal == 10)
    quant verbal train
1     10     10     3
```

Above we asked R to filter out cases having quant and verbal scores equal to 10. Since only a single case in our data contained these parameters, it filtered out only one row of data. We haven't named the object anything yet, we've only requested the operation to filter. We can easily name the object:

```
> iq.train.q10.v10 <- filter(iq.train, quant == 10, verbal == 10)
> iq.train.q10.v10
    quant verbal train
1     10     10     3
```

Notice above we had to call the new object so that it prints. That is, we had to specifically code iq.train.q10.v10 so that the new object is produced. In generating the object, we can both save it as well as print it by enclosing the entire assignment operation in parentheses, as in the following, which will simultaneously produce the new object as well as print it (give it a try, and you will see it prints the object):

```
> (iq.train.q10.v10 <- filter(iq.train, quant == 10,
verbal == 10))
```

2.10 How R Deals with Missing Data

Missing data is a fact of life in research, and whether one should replace missing values or not is a deep topic. For now, we are simply interested in learning how we can communicate to R that a value is missing. The following creates a vector, where we identify the last value of the vector as **NA** or "not available":

```
> v <- c(1:9, NA)
> v
 [1]  1  2  3  4  5  6  7  8  9 NA
```

We can use is.na() to tell us for which values of the vector there are missing values:

```
> object <- is.na(v)
> object
 [1] FALSE FALSE FALSE FALSE FALSE FALSE FALSE FALSE FALSE
TRUE
```

That is, the above reads that for the first 9 values, there are no missing values (i.e. missing values is FALSE). For the 10th value in our data, there is a missing value (i.e. missing values TRUE). If we tried to compute a mean on v, we would get:

```
> v <- c(1, 2, 3, 4, 5, 6, 7, 8, 9, NA)
> mean(v)
[1] NA
```

The way to get around this is to use na.rm = T, which removes the missing value. We get a mean of values 1 through 9, ignoring the missing element in position 10:

```
> mean(v, na.rm=T)
[1] 5
```

Sometimes, R will produce what amounts to an error message if we compute something that is not a number. It will use NaN to tell us "not a number":

```
> 0/0
[1] NaN
```

However, is.na() will generate a TRUE value for both NA and NaN values.

2.11 Using ls ( ) to See Objects in the Workspace

The ls() function will allow us to observe all the current objects in R's environment:

```
> ls()
 [1] "Arbuthnot"    "bindat"      "c"            "CanCor"
 [5] "ChestSizes"   "Dactyl"      "DrinksWages"  "Fingerprints"
 [9] "fit"          "fit.qv"      "fit.qv.int"   "GaltonFamilies"
[13] "Jevons"       "logit.fit"   "Macdonell"    "Nightingale"
[17] "Prostitutes"  "q"           "qv.data"      "rho"
[21] "train"        "v"           "x"            "y"
[25] "Yeast"
```

Notice that for our current session, we have many objects, which include "x" and "y" toward the end of it. We can use the function rm() to remove objects. Let's remove these two objects:

```
> rm(x, y)
> ls()
 [1] "Arbuthnot"    "bindat"      "c"            "CanCor"
 [5] "ChestSizes"   "Dactyl"      "DrinksWages"  "Fingerprints"
 [9] "fit"          "fit.qv"      "fit.qv.int"   "GaltonFamilies"
[13] "Jevons"       "logit.fit"   "Macdonell"    "Nightingale"
[17] "Prostitutes"  "q"           "qv.data"      "rho"
[21] "train"        "v"           "Yeast"
```

Notice that "x" and "y" are no longer listed. If we chose to remove all objects at the same time, we could use rm(list=ls():

```
> rm(list=ls())
> ls()
character(0)
```

After implementing the function, `ls()` generates `character(0)`, meaning we have no objects in the list.

Now that we have removed x from the workspace, let's see what happens when we try computing a mean for x:

```
> mean(x)
Error in mean(x) : object 'x' not found
```

Of course, since we removed x, R tells us the object is not found.

To quit R, simply code `q()`. Upon quitting, it will ask you whether you wish to save the workplace image. If you choose to save, R will create a file with extension `.RData` that will be saved to your current working directory. If you choose not to save, you will lose the information in your current workspace, and upon initiating R again, will start with a fresh workspace.

R allows you to get help on many functions. To get help, simply code `help()` where one would enter in parentheses the name of the function.

One can request a bunch of mathematical constants built into R, such as pi or e. For example:

```
> pi
[1]  3.141593
> print(pi)
[1]  3.141593
```

To create a sequence of numbers in R, for instance, from 1 to 10, we code:

```
> 1:10
 [1]   1  2  3  4  5  6  7  8  9 10
```

We can also specify not only the range of numbers for a sequence, but also the requested interval between numbers:

```
> seq(from=1, to=100, by=5)
 [1]   1  6 11 16 21 26 31 36 41 46 51 56 61 66 71 76 81 86
91 96
```

The above says to start at 1, go to 100, but by intervals of 5. To get a series of repeated values, we can code:

```
> rep(10, times = 100)
 [1] 10 10 10 10 10 10 10 10 10 10 10 10 10 10 10 10 10 10 10 10 10 10 10 10 10
[26] 10 10 10 10 10 10 10 10 10 10 10 10 10 10 10 10 10 10 10 10 10 10 10 10 10
[51] 10 10 10 10 10 10 10 10 10 10 10 10 10 10 10 10 10 10 10 10 10 10 10 10 10
[76] 10 10 10 10 10 10 10 10 10 10 10 10 10 10 10 10 10 10 10 10 10 10 10 10 10
```

The above told R to repeat the number 10 a total of 100 times.

2.12   Writing Your Own Functions

You can create your own functions to compute things in R using the `function()` statement. For example, the following creates a function to compute a simple arithmetic mean:

```
> m <- function(x) mean(x)
> m
function(x) mean(x)
```

Now at first glance, it doesn't appear as though we accomplished much, and simply generated something that prints out text. However, when we now include numbers, then request the function m, R will know what to do with it (this is left as an exercise).

2.13   Writing Scripts

To execute a series of lines at once instead of having to re-type each time, you can open a separate script window as follows. Select: `File -> New Script`, which will generate the following empty window (left):

We can then type in a series of codes in that window, such as appears on the right, where we reproduced the correlation matrix we entered earlier, and also computed the inverse of it. The difference between the code earlier and this code, is that using the script, we can enter everything at once, highlight, and execute it. Those coming from SAS for example will find this type of running code very familiar. Running code this way also allows one to visualize the entire program, so you can fine-tune it and make adjustments before running the series of commands. You can also view your command history by `history()`, or specify the most recent lines of history by, for instance, `history(50)`. Give it a try.

2.14 How to Create Factors in R

Often, we need to convert a variable entered in numerical form to a **factor**. For instance, when conducting an ANOVA, we typically have to do this so that R recognizes the variable as a factor with levels and not a variable measured on a continuous scale. Generating the factor can be done "on the spot" in the ANOVA code, or, beforehand to prepare the variable for the ANOVA. Let's look at an example:

```
> v <- c(0, 1, 2, 3, 4)
> v
[1] 0 1 2 3 4

> v.f <- factor(v)
> v.f
[1] 0 1 2 3 4
Levels: 0 1 2 3 4
```

Notice that for this example, we get the same values for the factor variable as we did the original variable. However, it needs to be emphasized that R is not treating these two things the same way. That is, before we converted v to a factor, we could compute a mean:

```
> mean(v)
[1] 2
```

However, when we try to compute a mean on the newly created factor variable v.f, we get the following:

```
> mean(v.f)
[1] NA
Warning message:
In mean.default(v.f) : argument is not numeric or logical:
returning NA
```

We can also ask R to confirm for us that v.f is now a factor, and that v is not:

```
> is.factor(v.f)
[1] TRUE
> is.factor(v)
[1] FALSE
```

Note that if we request the levels of the factor versus the original variable, we get:

```
> levels(v.f)
[1] "0" "1" "2" "3" "4"
> levels(v)
NULL
```

We can confirm that v.f has 5 levels, and v has none (since it is not a factor, it does not have levels):

```
> nlevels(v.f)
[1] 5
> nlevels(v)
[1] 0
```

We could also name the new levels by different values, if we so chose. In our example, even though our original data has numbers 0, 1, 2, 3, 4, we may wish to name the levels differently. We can accomplish this via:

```
> v.f <- factor(v, levels = c(10, 11, 12, 13, 14))
> v.f
[1] <NA> <NA> <NA> <NA> <NA>
Levels: 10 11 12 13 14
```

We see that the new variable now has levels 10 through 14 instead of the original values of the variable 0 through 4.

There are of course many other functions available in R to compute basic statistics, though we do not demonstrate them here. Most of them are self-explanatory, and include the following: max(), min(), sum(), median(), range(), var(), cor(x,y), sort(), rank(), quantile(), cumsum(), cumprod(), fivenum(), where this last one generates a **five-number summary**, which includes the minimum, lower hinge, median, upper hinge, and maximum of a data set. Five-number summaries are used in the construction of boxplots, to be surveyed in the following chapter.

2.15 Using the table() Function

The table() function allows one to get a summary of the different values in a vector. For instance, consider the following vector. When we apply the table() function to it, it tells us there are two 0s, then each one value of 1, 2, 3:

```
> x <- c(0, 0, 1, 2, 3)
> table(x)

x
0 1 2 3
2 1 1 1
```

As one can imagine, the `table()` function is especially useful for very large data sets. Consider once more the data `GaltonFamilies` found in the **HistData** package, a data set containing observations for 934 children in 205 families on which Galton originally analyzed going back to 1886:

```
> attach(GaltonFamilies)
> some(GaltonFamilies)
      family father mother midparentHeight children childNum gender childHeight
165    045   71.0   65.0            70.60        3        1   male        68.0
237    061   70.0   69.0            72.26        4        1   male        71.0
415    097   69.0   68.5            71.49       10        3   male        70.0
553    123   69.5   61.0            67.69        5        5 female        62.0
596    133   68.0   65.5            69.37        7        6 female        65.2
665    145   68.0   63.0            68.02        8        1   male        71.0
699    155   68.0   60.0            66.40        7        4 female        60.0
717    158   68.0   59.0            65.86       10       10 female        61.0
881    190   65.0   65.0            67.60        9        8 female        61.0
902    196   65.5   63.0            66.77        4        1   male        71.0
```

We will use the `table()` function to generate the different frequencies associated with the variable `father`:

```
> table(father)
father
  62 62.5    63    64    65 65.5    66 66.5    67 67.5    68 68.2 68.5 68.7    69 69.2
   3    2     2    17    42    9    57   21    48   16   102    5   36    4   116    4
69.5 69.7    70 70.3 70.5    71 71.2 71.5 71.7    72 72.5 72.7    73 73.2    74 74.5
  37    1   131    7   38    91    2   11    5    42    9    8    25    1    20    1
  75 75.5 78.5
  13    4    4
```

We can see from the above that there are 3 values of 62, 2 values of 62.5, 2 values of 63, and so on. We can also use `tapply()` to get specific numerical summaries for categories of a factor:

```
> tapply(children, gender, mean)
  female      male
6.344371 6.008316
```

The above reports the mean of children for each category of gender.

2.16   Requesting a Demonstration Using the `example()` Function

You will definitely want to keep this hint as you work your way through functions throughout this book and others you encounter in your research career. When you come across a function, and are unsure on how to use it, you can request R to generate an example of how that function is used. This is accomplished by using the `example()` function. For instance, suppose you wanted to compute a mean, but were unsure of how this might look in R. You can compute the following to get a demonstration:

```
> example(mean)
mean> x <- c(0:10, 50)
mean> xm <- mean(x)
mean> c(xm, mean(x, trim = 0.10))
[1] 8.75 5.50
```

In the demonstration, R gives an example of computing a mean on a series of numbers. Or, if you wanted an example of how to specify a variable as a factor in ANOVA, you could request `example(factor)` to give you a demonstration.

2.17 Citing R in Publications

After you've used R for your statistical analyses, a great thing to do would be to cite it to give proper credit. For details on how to cite it, we simply code:

```
> citation()
To cite R in publications use:

  R Core Team (2018). R: A language and environment for statistical
  computing. R Foundation for Statistical Computing, Vienna, Austria.
  URL https://www.R-project.org/.

A BibTeX entry for LaTeX users is

  @Manual{,
    title = {R: A Language and Environment for Statistical Computing},
    author = {{R Core Team}},
    organization = {R Foundation for Statistical Computing},
    address = {Vienna, Austria},
    year = {2018},
    url = {https://www.R-project.org/},
  }

We have invested a lot of time and effort in creating R, please cite it
when using it for data analysis. See also 'citation("pkgname")' for
citing R packages.
```

Exercises

1   Generate a vector with elements 1 through 10 by coding `1:10`, then remove the tenth element of that vector. Then, use the function `any()` to confirm that the tenth element no longer exists. For this, code `any(x > 9)`.

2 Use the `rep()` function to create a vector 0, 1, 2 with each value repeating a total of 10 times. To obtain this, code the following, then change the number of repetitions from 10 to 5 by altering `each = 10` to `each = 5`:

```
> rep(c(0, 1, 2), each = 10)
```

3 Consider the following two vectors:

```
> x <- c(1, 2, 3, 4, 5)
> y <- c(1, 2, 3, 4, 6)
```

By coding `y > x`, confirm that the fifth element of vector `y` is greater than the fifth element of vector `x`.

4 Consider once more the vector `x`:

```
> x <- c(1, 2, 3, 4, 5)
```

Define a new vector `z`, such that it **filters** out squared elements of `x` greater than 10. To do this, define `z` as: `z <- x[x*x > 10]`. Confirm for yourself that the correct elements were retained in the new vector.

5 Using the `abs()` function, confirm that the absolute value of the number −2 is 2.

6 Consider the following vector `w <- c(0, 1, 9, 2, 5)`. Using the `sort()` function, sort the elements from smallest to largest, then use the `order()` function on the original vector. What might the `order()` function be telling you?

7 Evaluate whether the following two sets are equal using `setequal()`:

```
> x <- c(0, 1, 3)
> y <- c(0, 3, 1)
```

8 Use the `methods()` function to learn more of the generic functions of the `plot()` function. That is, code `methods(plot)` and report the output.

9 Demonstrate the computation of a mean m using the following function:

```
> m <- function(x) mean(x)
```

10 Obtain an example of the `median()` function using `example()`.

3

Exploring Data with R

Essential Graphics and Visualization

LEARNING OBJECTIVES

- How to use R's `plot()` function to produce basic plots.
- How to generate histograms, stem-and-leaf plots, boxplots, and other useful depictions of data in a single dimension.
- How to produce plots that go beyond boxplots and scatterplots such as violin plots, sunflower plots, and others.
- How to add a smoother and jittering to a plot.
- How to produce a Q–Q plot, and use it as a visual assessment of normality of a distribution.
- How to create and interpret multidimensional plots for dimensions of two (bivariate) or greater (multivariate).
- How to produce pie graphs.
- How to use R to produce tables to read the frequencies of various observations.

3.1 Statistics, R, and Visualization

In the early 1900s, the focus of statistics was more or less on **confirmatory inference**. It was then that such techniques as the **analysis of variance** and **null hypothesis testing** arose, and the priority was to obtain p-values to provide a sense of control and management over uncertainty. That is, if one could reject a null hypothesis at $p < 0.05$, though one may not be sure if the null was actually false, one could at least have a sense of the degree to which he or she might be wrong in rejecting the null. P-values and inference were "big deals" back then.

Inference is still extremely important today, but with the advent of fast computing power, the exploration of small to large "big data" data sets has never been easier. As a result of the ease with which data can be obtained today through computers and the internet, as discussed previously, it is not uncommon to have data sets in the range of billions to trillions of variables on huge amounts of individuals,

Univariate, Bivariate, and Multivariate Statistics Using R: Quantitative Tools for Data Analysis and Data Science, First Edition. Daniel J. Denis.
© 2020 John Wiley & Sons, Inc. Published 2020 by John Wiley & Sons, Inc.

especially in fields such as business, marketing, and social media. Such data sets were generally not possible or available when much of mathematical statistics were developed in the early 1900s, but even if they were, the facility to graph such data would have been impossible without the massive computing power we have today. What would have taken forever back then to graph, today such plots and visualization can be obtained in a matter of seconds with even elementary computing skills.

Today, a data analyst or data scientist can obtain a huge data set and extract visual patterns from that data in a very short time, allowing him or her to quickly gain insights into the data that would have been impossible to obtain without the modern availability of computing power. R software is especially good at generating high-level graphics, largely due to its flexibility in allowing the user to program exactly what they wish to see from a graphic.

In this chapter, we survey a few of the graphing possibilities in R, from **basic** to more **advanced**, and feature many visualization tools that you can immediately apply directly to your data as needed. We only really have space to simply scratch the surface in this regard, but our treatment will be enough to give you that "foot in the door" as a foundation for exploring further options as you may need them. We also use some of the plots surveyed here in other places in the book to portray output from individual analyses. "A picture is worth a thousand words" is definitely true, and graphs can help you tell a story about your data that would otherwise go untold with only numerical summaries and statistics. While inferential statistics and effect size measures can help us draw relatively reliable conclusions from our data, graphs and visualizations can help make the scientific findings accessible to virtually anyone, even with minimal coursework in statistics or data science. Hence, if you are working on a presentation for a supervisor, boss, or any other authority, a strong graphic can go a long way to showing him or her what the data are "saying" without getting too bogged down in technicalities.

What graphic or visualization technique one employs for any particular situation often comes down to individual preference and taste, but the good thing about R is that there is no shortage of possibilities to choose from. Thanks to advanced packages such as **ggplot2** and others, continuous improvements and advances to R's graphics are being made daily. We will briefly discuss ggplot2 a bit later in the chapter, as it is one of the most elite graphing packages available in R. You can load ggplot2 as a separate package, or you can also load a more global package called **tidyverse** that contains ggplot2 as well as a few other packages useful for data analysis in R:

```
> install.packages("tidyverse")
```

To make it available for use, we code:

```
> library(tidyverse)
```

Because of the aforementioned availability of computing power, it is definitely an exciting time to be specializing in the field of computer graphics and visual displays, so if this is an area of interest for you, you are strongly encouraged to give it hot pursuit! The future of data analysis and data science most likely belongs to those who can communicate quantitative findings effectively using advanced visualization.

3.2   R's plot() Function

The most common way to obtain a basic plot in R is to use the plot() function. R is **polymorphic**, meaning that a single function can usually be applied to a whole host of **objects**. As we will see throughout the book, plot() is one such function that we can apply to various objects and each time will generate slightly different output depending on the object. As an example, we obtain a plot of our previously defined vectors x and y. For convenience, we reproduce the vectors here:

```
> x <- c(1, 3, 5, 7, 9)
> y <- c(2, 4, 6, 8, 10)
> plot(x, y)
```

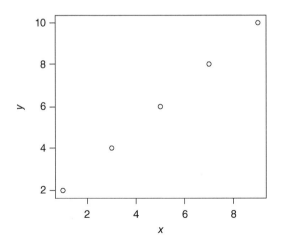

We will see throughout the book how we can add to the plot() function to improve our graphics, but for now, we simply consider the basic plot above. That the plot exhibits linearity is no surprise, given the nature of our vectors (i.e. x is a vector of odd numbers, y is a vector of even numbers). Otherwise, the plot is unin-teresting, as there are very few data points. We can join the dots in the plot using `lines()`:

```
> lines(x, y)
```

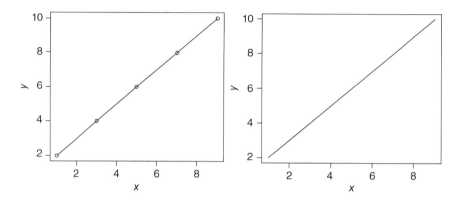

If we wanted the line plotted, but did not want to see the points it is joining, we could have used `plot(x, y, type = "l")` where `type` is defined with the letter "l" to accomplish this (right-most plot above). Instead of a simple plot as above, let's instead obtain a plot of random numbers. We will use the function `rnorm()` to define two vectors generating 1000 random observations in each vector. We will then plot the resulting distributions:

```
> z = rnorm(1000)
> w = rnorm(1000)
```

The first line of code requests R to produce a list of 1000 random numbers, and store these numbers in the object z. The second line of code likewise requests R to store 1000 random numbers in a different object, this time w. We use the **car** package in order to employ the function `some()` to generate a few of the observations of the 1000 generated from each vector:

```
> library(car)
> some(z)
[1]  -1.22010109   0.09990338  -0.53525670  -0.94100974
     -1.41989896   2.01933254
[7]  -0.08626850  -0.73251531   0.80543729  -3.02230321

> some(w)
[1]  -0.9355866   0.5552930  -0.7325919  -0.3101065
      2.6748709  -0.4663131
[7]   1.8991274   0.8067137   0.4540569   1.6397328
```

Plotting z and adding a `lines ()` function to it can give us a visual sense of the variability in the plot:

```
> plot(z)
> lines(z)
```

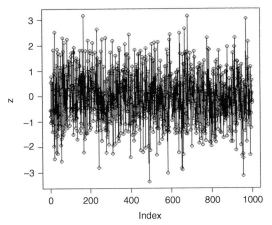

Each small circle represents a data point in the plot, depicting much variability around z = 0. We can produce a **histogram** of z, also employing what is known as a **rug plot**, which can help communicate density of points along the *x*-axis:

```
> hist(z)
> rug(z)
```

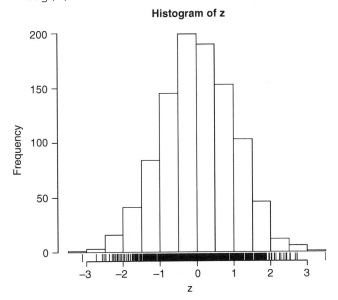

The reason why it is called a "rug plot" is because it adds a layer at the bottom of the histogram representing the distribution of points. This can help communicate density. If you look at the bottom of the plot, you will see that most of the density of the plot occurs near the center around 0, and less toward the extremes as we move toward −3 and +3. A bit later we will see how to communicate density in a bivariate setting with sunflower plots.

We load the package **HistData** to demonstrate another histogram. We will use the HistData package to demonstrate a variety of analyses throughout the book. The package consists of a number of relatively small data sets that are especially meaningful historically from a visualization point of view. The first data set below is Galton's data, on which midparent height (the mean of heights of both parents, simply referred to as parent in the data) and that of child height (the height of the offspring as a grown adult) were recorded:

```
> library(HistData)
> some(Galton)
     parent child
55     69.5  64.2
133    67.5  65.2
222    67.5  66.2
228    67.5  66.2
377    66.5  67.2
419    70.5  68.2
594    68.5  69.2
719    69.5  70.2
735    69.5  70.2
809    69.5  71.2
```

The full data set consists of 928 observations. Below we obtain a histogram of parent height:

```
> attach(Galton)
> hist(parent)
```

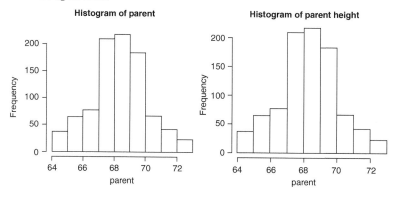

We can see that the parent variable follows an approximately normal distribution (left plot). We adjust the title of the plot (> hist(parent, main = "Histogram of parent height"))(right plot). These data are plotted univariately because they feature only a single variable. Shortly, we will plot data on parent height and child height simultaneously in the well-known scatterplot, and also see how we can extend on the scatterplot to create more elaborate bivariate plots.

3.3 Scatterplots and Depicting Data in Two or More Dimensions

Having surveyed a few ways to show data in a single dimension, often we would like to plot two variables against each other in **two dimensions**. That is, we would like to see scores on both variables **bivariately** rather than univariately. We begin with a generic example using vectors z and w, then proceed to produce additional scatterplots using our featured data sets.

As a generic example then, recall how we generated data on z and w:

```
> z = rnorm(1000)
> w = rnorm(1000)
```

We now obtain a plot of z and w to reveal their bivariate (as opposed to univariate) distribution:

```
> plot(w, z)
```

Make sure you understand what each point in the plot represents. The points no longer represent only scores on a single variable as in the previous univariate plots.

Rather, each point represents a **joint observation** on w and z. The axes are automatically labeled "z" for the *y*-axis, and "w" for the *x*-axis. We can make them more interesting by giving them names via xlab and ylab statements, as well as adjust pch, which is the **plotting character**:

```
> plot(w, z, main = "Plot of W and Z",
+ xlab="z for absicca axis", ylab="w for ordinate axis",
pch = 19)
```

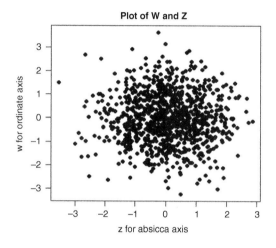

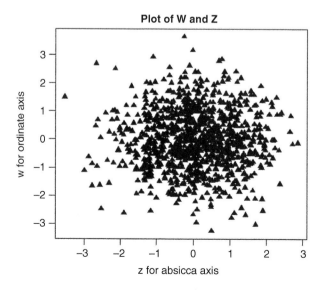

The pch = 19 code stands for "plotting character." A number of different plotting characters are possible, some of which are given below:

0	1	2	3	4
□	○	△	+	×

5	6	7	8	9
◇	▽	⊠	✳	⊕

10	11	12	13	14
⊕	⧓	⊞	⊠	◸

15	16	17	18	19
■	●	▲	◆	●

20	21	22	23	24	25
•	●	▪	◆	▲	▼

To demonstrate, instead of using pch = 19, we try pch = 17 (second plot above). What symbol one uses is a matter of taste and individual preference, but it is clear that R offers plenty of options in this regard. The points () function can also be used to embellish a plot (e.g. try points (w, z, pch="+")).

3.4 Communicating Density in a Plot

The concept of **density** is prevalent in virtually all data sets other than perhaps on very simple ones. We saw earlier how a rug plot can be used to communicate density on the *x*-axis in the univariate setting. In a bivariate plot, we would also like to communicate density, but this time the density should refer to the bivariate distribution rather than only univariate. For example, consider again a bivariate plot of Galton's data:

```
> plot(child, parent)
```

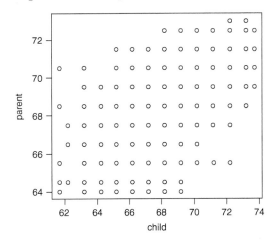

The above plot does not tell the full story of the relationship between parent and child, as it does not give any indication of density of points in the plot. At first glance, by looking at the plot one might assume an equal distribution of points. However, in the actual data, there is much more density toward the center of the distribution than in the outer edges, but the classic scatterplot above is unable to capture or communicate this density. Instead of such a traditional plot then, we can produce what is known as a **sunflower plot** that will capture the density of points in the bivariate distribution:

```
> sunflowerplot(parent ~ child)
```

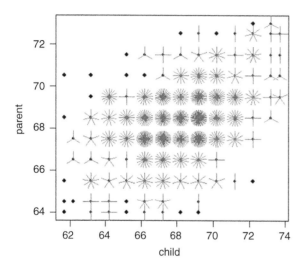

Each "sunflower" in the plot, ranging from only a single point in some places to true multi-leaf sunflowers containing several leaves, depicts the number of points in a particular bivariate "bin" containing a range of scores on child and parent. For example, we can see from the above plot that the dense sunflowers are toward the center, which communicates that these bins contain frequencies of several scores. Bins on the outer edge contours of the plot for our data contain fewer points. The sunflower plot then is excellent if one wants a quick and easy way to inspect the overall density of a bivariate scatterplot. That is, it communicates more than a bivariate distribution of points, in that it also gives an indication of the density of points via "overplotting" in different areas of the plot.

Referring back to our original plot, note that we could have also obtained the plot by requesting to extract the specific variables from the Galton data frame, where `Galton$parent` tells R to pull out the `parent` variable from the Galton data frame, and `Galton$child` tells R to pull out the `child` variable from the same data frame:

```
> plot(Galton$parent, Galton$child)
```

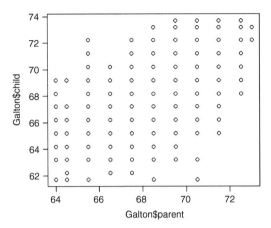

However, notice that using this approach, R automatically titles the *x*- and *y*-axes as `Galton$parent` and `Galton$child`, respectively.

Sometimes we wish to perform transformations on variables while plotting them. Using `qplot()` in the **ggplot2** package, we can accomplish this very easily. For example, let's plot the log of `parent` against the log of `child`:

```
> qplot(log(parent), log(child))
```

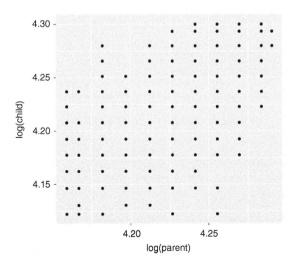

Using qplot(), we can adjust the transparency of the points to shade out observations toward the outer margins of the plot. This is done by adjusting the **alpha aesthetic**:

```
> qplot(log(parent), log(child), alpha = I(1/10))
```

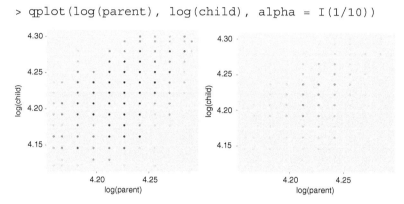

If we adjust alpha = I(1/10) to I(1/100), an even smaller fraction, we will effectively shade out observations that are marginal to the center of the plot: > qplot(log(parent), log(child), alpha = I(1/100)) (see Wickham, 2009, p. 73 for more specifics on exactly how this function works and its relation to overplotting).

A **smoother** can easily be added to the plot as well. The idea of "smoothing" a plot is to help visualize the trend of the data:

```
> qplot(log(parent), log(child), geom = c("point",
"smooth"))
`geom_smooth()` using method = 'loess' and formula 'y ~ x'
```

We can see in the above plot that the shaded area around the curve increases slightly as we move to the extremes of the plot. Smoothing can also be performed by a grouping variable. For example, the following is a plot of `childHeight` by `midparentHeight`, but for each gender (male vs. female) separately (you will need to produce the graph yourself to see the color variation):

```
> qplot(childHeight, midparentHeight, data = GaltonFamilies,
color = factor(gender),
+ geom=c("point", "smooth"))

`geom_smooth()` using method = 'loess' and formula 'y ~ x'
```

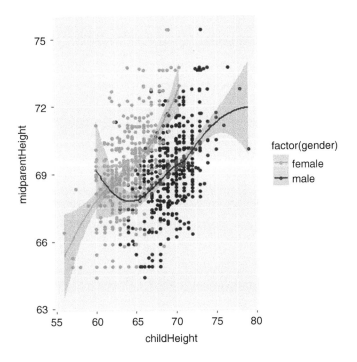

Jittering in a plot is another way to reveal density. In the following, we plot the log of `parent` and `child` using `qplot()`, and specify the **geom** as "jitter" to tell R to provide an indication of density within each bivariate "cell" in our plot. That is, for overlapping points, R will produce a slight scatter of "noise" that depicts density:

```
> qplot(log(parent), log(child), geom = "jitter")
```

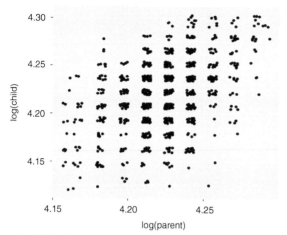

Bubble charts can also at times be useful, where this time, the size of the "bubbles" corresponds to the variable `childNum` in the example below, representing the birth order of the given child (again, you'll need to produce the graphic yourself to see the color variation by gender):

```
> attach(GaltonFamilies)
> g <- ggplot(GaltonFamilies, aes(childHeight, midparentHeight))
> g + geom_jitter(aes(col = gender, size = childNum))
```

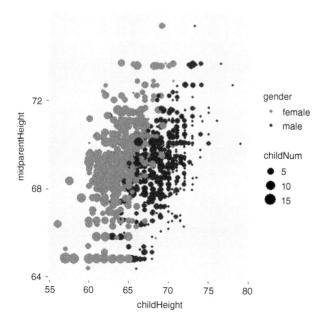

3.5 Stem-and-Leaf Plots

We can easily generate in R what are known as **stem-and-leaf plots**. These are akin to "naked histograms" that are tipped over on their side, and have the advantage that one can see the actual data points in the plot rather than the data be invisibly binned as in the case of histograms. To demonstrate a stem-and-leaf, we will use the `faithful` data, part of the base package in R, a data set featuring both the waiting time between eruptions and the duration of each eruption for the Old Faithful geyser at Yellowstone National Park in Wyoming. The following depicts the distribution of eruption time in minutes for a given eruption:

```
> attach(faithful)
> stem(eruptions)

  The decimal point is 1 digit(s) to the left of the |

  16 | 070355555588
  18 | 000022233333335577777777888822335777888
  20 | 00002223378800035778
  22 | 0002335578023578
  24 | 00228
  26 | 23
  28 | 080
  30 | 7
  32 | 2337
  34 | 250077
  36 | 0000823577
  38 | 2333335582225577
  40 | 000000335778888800223355557778
  42 | 0333555577880023333555577778
  44 | 022223555778000000002333357778888
  46 | 000023335770000023578
  48 | 00000022335800333
  50 | 0370
```

To help understand what the plot is communicating, let's obtain a frequency table of this data using the `table()` function (partial table shown below). We see that there is one value of 1.6 in the data, one value of 1.667, one value of 1.7, 6 values of 1.75, and so on. But how are these represented in the plot? The first value of 1.6 is represented by the first "0" to the right of "|" in the first line. Since, as R tells us, The decimal point is 1 digit(s) to the left of the |, the

"16" in the plot to the left of "|" is understood to mean "1.6" with the first digit to the right of "|" corresponding to "0," so the first data point is 1.60. The next data point is rounded up from 1.667 to 1.67 in the plot (i.e. the second digit to the right of "|"). At this point, the plot can seem confusing, because the next digit just next to the "7" is another "0." However, what is implied here, is that we have moved to a stem of 1.7, and so that "0" corresponds to the number 1.70 in our data. The next number "3" corresponds to a single 1.73 in our data, then we have 6 values of 1.75, which confirms what we see in the data produced by the `table()` function (again, we table only the first few cases for demonstration):

```
> head(table(eruptions))
eruptions
  1.6 1.667    1.7 1.733    1.75 1.783
    1     1      1     1       6     2
```

Hence, that first line of the stem plot consists not only of data with 1.6 as the stem, but also 1.7. That is, it is an interval of data points within the range 1.6–1.7. We will see how we can adjust this interval in a moment.

For the next line in the stem plot, the stem is 1.8–1.9. The numbers in that second line of the plot correspond to the following, which you can readily verify in the table generated below:

```
1.80,  1.80,  1.80,  1.80
1.82,  1.82,  1.82
1.83,  1.83,  1.83,  1.83,  1.83,  1.83,  1.83
1.85,  1.85
1.87,  1.87,  1.87,  1.87,  1.87,  1.87,  1.87,  1.87
1.88,  1.88,  1.88,  1.88
1.92,  1.92
1.93,  1.93
1.95
1.97,  1.97,  1.97
1.98,  1.98,  1.98
```

Now, if we find the stem intervals too wide, we can adjust them. The default in the above plot was set at `scale = 1` (we didn't have to explicitly state this in the R code), but if we adjust to `scale = 2`, we will no longer have to try to figure out what is happening within each line of the plot, as R will start each stem at a value of 1.6, 1.7, etc. We print only a few of the observations to give you an idea:

```
> stem(eruptions, scale = 2)
```

The decimal point is 1 digit(s) to the left of the |

```
16 |  07
17 |  0355555588
18 |  0000222333333355777777778888
```

The disadvantage of plotting data this way is that the plot will be much wider, which though providing more information within each line, may not provide as much of an overall summary or shape as we may wish. Likewise, setting the scale too small may provide too "condensed" of a summary:

```
> stem(eruptions, scale = 0.5)

  The decimal point is at the |

1 |  67778888888888888888888888999999999999999999
2 |  0000000000000000011111112222222223333333344444444
2 |  566899
3 |  133344
3 |  55566666777888888888999999
4 |  00000000000011111111111111122222222222223333333333
     3333334444444444444
4 |  5555555555555555555566666666666666666667777777777788888
     888888889999999
5 |  0011
```

In the end, how we set scales in graphs and how we choose to **bin data** is a matter of taste and preference. However, always recall that a good graphic is supposed to communicate the data as objectively as possible with a sense of fairness and ethics so that the graph is not purposely misleading.

3.6 Assessing Normality

Normality of distributions is an assumption required of many popular statistical methods. **Q–Q plots** are a graphical way of assessing normality. These plots compare actual values of distributions with theoretical values that would be expected if the distribution were exactly normal. For a perfectly normal distribution, all points should line up along the line spanning from the bottom left to the top right. The extent to which points deviate from the line is indicative of deviation from normality. For our z data, we can see that since the data more or less fall along the line, the distribution is relatively normal in shape.

```
> qqnorm(z); qqline(z)
```

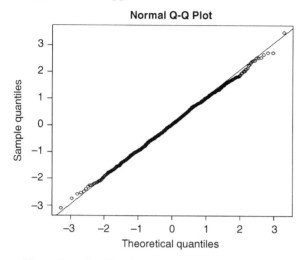

Normal Q-Q Plot

Normality of a distribution can also be assessed inferentially using a variety of tests. Recall that by "inferential" here, we mean the test will evaluate a null hypothesis that the data are normally distributed in the population from which the sample data were drawn. One such test available in R is the **Shapiro–Wilk test**. We demonstrate it first on the z data:

```
> shapiro.test(z)

        Shapiro-Wilk normality test

data:   z
W = 0.99925, p-value = 0.9681
```

The null hypothesis is that the data are normally distributed. Since p is quite high (definitely not less than some preset level of significance such as 0.05 or 0.01), we fail to reject the null hypothesis. That is, we have insufficient evidence to suggest the distribution of z is not normally distributed. Hence we may proceed with the assumption that the data are more or less normal (we haven't confirmed the assumption, we simply do not have evidence to reject it). For the eruptions data, as one might suspect, the result is quite different (plot a histogram to see why):

```
> shapiro.test(eruptions)

        Shapiro-Wilk normality test
```

```
data:   eruptions
W = 0.84592, p-value = 9.036e-16
```

Since the *p*-value is so low (recall that 9.036e−16 means to move the decimal point 16 places to the left), we reject the assumption of normality, and conclude the distribution is not normal. This of course agrees with the graphical evidence obtained earlier that the eruptions data are highly abnormal in shape. While the Shapiro–Wilk normality test is sometimes useful, usually simply a visual depiction of the data is sufficient to get a general assessment of normality using the afore-mentioned graphical devices. As we know already, *p*-values are sensitive to such things as sample size, and thus the Shapiro–Wilk test will likely reject as sample size increases. This is another reason why visual depictions are often deemed sufficient for drawing approximate conclusions about normality and usually formal inferential tests here are not needed.

3.7 Box-and-Whisker Plots

Box-and-whisker plots can be useful devices for visualizing a distribution around the **median** rather than the mean. A description of the box-and-whisker plot, and how to interpret one is given in **Box 3.1**. We generate a couple of options for box-plots in R below, the second using the **lattice** package. The first features a plot of parent height, the second a plot of child height:

```
> boxplot(parent, main = "Boxplot of Parent Height")
> library(lattice)
> bwplot(child, main = "Boxplot of Child Height")
```

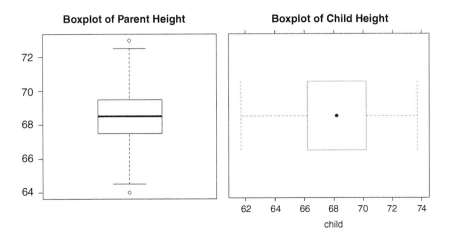

Box 3.1 How to read a box-and-whisker plot (Denis, 2019).

- The median in the plot is the point that divides the distribution into two equal halves. That is, 1/2 of observations will lay below the median, while 1/2 of observations will lay above the median.
- Q1 and Q3 represent the 25th and 75th percentiles, respectively. Note that the median is often referred to as Q2, and corresponds to the 50th percentile.
- IQR corresponds to "Interquartile Range" and is computed by Q3 − Q1. The semi-interquartile range (not shown) is computed by dividing this difference in half (i.e. (Q3 − Q1)/2).
- On the leftmost of the plot is **Q1 − 1.5 × IQR**. This corresponds to the lower-most "inner fence." Observations that are smaller than this fence (i.e. beyond the fence, greater negative values) may be considered to be candidates for outliers. The area beyond the fence to the left corresponds to a very small proportion of cases in a normal distribution.
- On the rightmost of the plot is **Q3 + 1.5 × IQR**. This corresponds to the upper-most "inner fence." Observations that are larger than this fence (i.e. beyond the fence) may be considered to be candidates for outliers. The area beyond the fence to the right corresponds to a very small proportion of cases in a normal distribution.
- The "whiskers" in the plot (i.e. the vertical lines from the quartiles to the fences) will not typically extend as far as they do in this current plot. Rather, they will extend as far as there is a score in our data set on the inside of the inner fence (which explains why some whiskers can be very short). This helps give an idea as to how compact is the distribution on each side.

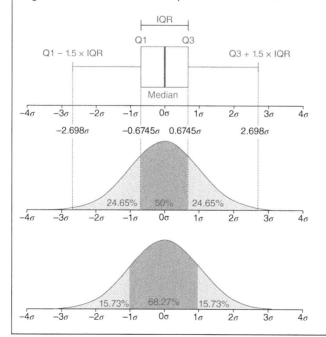

Using simply the `plot()` function, we can generate boxplots for each level of the factor variable. For example, in the following, `childHeight` is the response variable, while `gender` is the factor:

```
> attach(GaltonFamilies)
> plot(gender, childHeight)
```

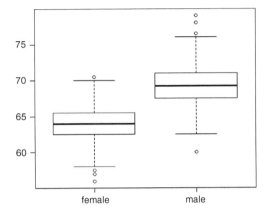

A quick glance at the plot reveals that the median for males is higher than the median for females, but that each distribution features a few extreme points, as indicated by circles beyond what are called the whiskers of the plots.

We can also use the **ggplot2** library to generate boxplots that reveal the actual data points within each plot:

```
> qplot(x = gender, y = childHeight, data =
GaltonFamilies, geom=c("boxplot", "jitter"), fill=gender)
```

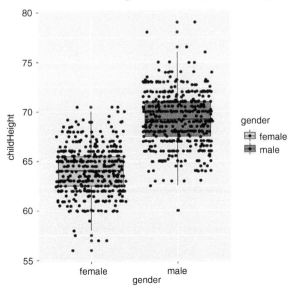

As mentioned, the library **ggplot2** is one of R's most prominent and valuable packages when it comes to visualization, and we use the package in many places throughout the book. **ggplot2** allows one to build graphics in a **layered fashion**, and the user can add to the plot's further specification as desired. For example, consider how we constructed the above boxplots. The coding of geom=c ("box-plot", "jitter") stands for "geometric object," for which in this case, a boxplot was chosen, along with a request to **jitter** data points. Recall that the jittering helps to reveal **density** of points in the plot. However, we could have selected a different **geom**, for instance, geom = "dotplot" would have produced:

```
> qplot(x = gender, y = childHeight, data =
GaltonFamilies, geom=c("dotplot", "jitter"), fill=gender,
binwidth = c(0.02, 200))
```

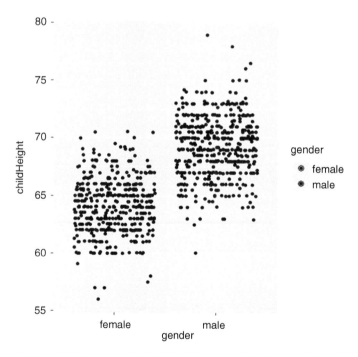

The point, for now, is that the advantage of using a package like **ggplot2** is that it is **flexible**, and whatever your graphical needs, it's likely that with some digging and experimenting, you can obtain the graphic you want or need for your situation. Do not expect immediate reinforcement as one would experience in SPSS, for instance. Generating graphics in R can be very time-consuming and can require much trial and error before you get the "perfect" plot you have in mind.

The **dot and box plot** is also useful for visualizing the distribution of points in a boxplot (code adopted from Holtz, 2018, *The R Graph Gallery*). Consider the following plot for a fictional data set called the achievement data which we will feature in the following chapter on ANOVA. It is a hypothetical data set on which achievement scores are hypothesized to be a function of teacher and textbook:

```
> g <- ggplot(achiev, aes(f.teach, ac))
> g + geom_boxplot() +
+ geom_dotplot(binaxis='y',
+ statckdir = 'center',
+ dotsize = .5,
+ fill = "red")
```

```
> achiev
   ac teach text
1  70     1    1
2  67     1    1
3  65     1    1
4  75     1    2
5  76     1    2
6  73     1    2
7  69     2    1
8  68     2    1
9  70     2    1
10 76     2    2
11 77     2    2
12 75     2    2
13 85     3    1
14 86     3    1
15 85     3    1
16 76     3    2
17 75     3    2
18 73     3    2
19 95     4    1
20 94     4    1
21 89     4    1
22 94     4    2
23 93     4    2
24 91     4    2
```

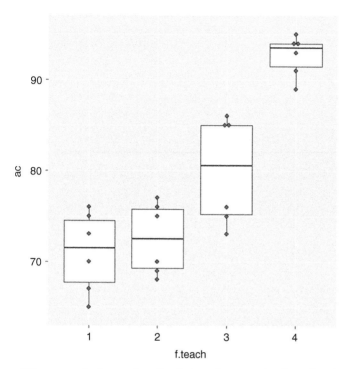

We can see by inspecting the data and comparing it to the plot that where there are two horizontal points in the boxplot, this corresponds to duplicates in the achievement data. For example, for teacher 4, there are two values of 94 (high-lighted in the data), as indicated by two periods placed horizontally in the plot. One must be cautious when generating dot and box plots, as if there are numerous values in the data that are equal, one could end up with a very confusing plot, as the following demonstrates on the **GaltonFamilies** data. While some information can be gleamed in terms of spotting the more extreme scores, and how many of those scores exist, the mass of data toward the center is far too "busy" for the plot to be of much use:

```
g <- ggplot(GaltonFamilies, aes(gender, children))
g + geom_boxplot() +
geom_dotplot(binaxis='y',
stackdir='center',
dotsize = .5,
fill="red")
```

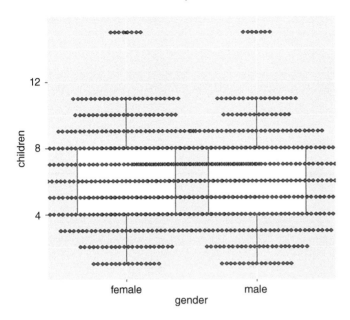

3.8   Violin Plots

A **violin plot** is sometimes a preferred alternative to boxplots for visualizing a distribution. It represents similar information as a boxplot, but is more accurate in displaying the density associated with a distribution. We feature an example of a few such plots for the **iris** data. Next to each are the comparative boxplots. To create the violin plot, we first convert the variable Species to a factor:

```
> iris$Species <- as.factor(iris$Species)
> head(iris)
  Sepal.Length Sepal.Width Petal.Length Petal.Width Species
1          5.1         3.5          1.4         0.2  setosa
2          4.9         3.0          1.4         0.2  setosa
3          4.7         3.2          1.3         0.2  setosa
4          4.6         3.1          1.5         0.2  setosa
5          5.0         3.6          1.4         0.2  setosa
6          5.4         3.9          1.7         0.4  setosa
```

To obtain a violin plot, we will rely again on the package **ggplot2**, and will choose the geom to be that of "violin":

```
> library(ggplot2)
> iris.violin <- ggplot(iris, aes(x = Species, y=Sepal.
Length)) + geom_violin()
> iris.violin
```

```
> iris.boxplot <- ggplot(iris, aes(x = Species, y=Sepal.
Length)) + geom_boxplot()
> iris.boxplot
```

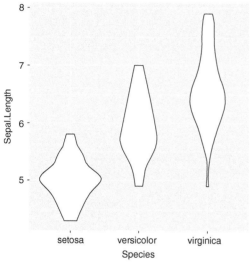

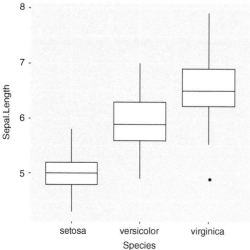

Note that the advantage of violin plots over ordinary boxplots is that they contain a measure of the density of the distribution represented by its width. For example, comparing the violin plot of setosa to that of the corresponding boxplot, we can see that the violin plot is providing more "information" about the distribution because it reveals a measure of horizontal width of probability density, whereas the boxplot does not. For `setosa`, we can see there is much density toward the center of the distribution, whereas for `versicolor`, the density is primarily toward one end of the distribution rather than in the center. Sometimes violin plots are a preferred alternative to boxplots, or can be used in conjunction with boxplots to reveal more about the distribution. For more details and examples of violin plots and how to read and interpret a variety of distributions, see Hintze and Nelson (1998).

3.9 Pie Graphs and Charts

In the world of graphics and visualization, graphing experts (i.e. those who specialize primarily in inventing visualization tools) for the most part generally recommend against **pie graphs**, and instead typically recommend tools that have a better **horizon** than do pie graphs. That is, one of the problems with pie graphs is that it can be relatively difficult to detect which category has more **area** in a given plot because unlike with a histogram or similar plot, for instance, one cannot easily detect a higher bin as one looks toward the horizon (imagine a flat landscape and trying to conclude which tree is taller). As a simple example, consider the following data on vector x:

```
> x <- c(1, 2, 3, 3)
```

Suppose we use a **barplot** to visualize the distribution, then compare it side-by-side to a pie graph:

```
> barplot(x)
> pie(x)
```

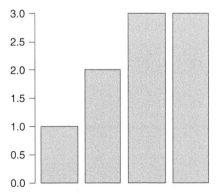
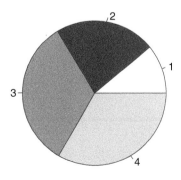

Notice that immediately, it is apparent in the barplot that two categories have the same value since we can tell by the horizon that the last two categories have equal height. Can we tell this from the pie graph? Not so easily. That is, if we did not know in advance based on the actual data, it would be very challenging to conclude from the pie graph that categories 3 and 4 have equal frequencies, similar to how it can be terribly difficult to know who is getting more pumpkin pie at Thanksgiving dinner! Now, having said this, and despite what graphing experts warn us about using them, the pie graph is popular and is likely going nowhere anytime soon. Researchers still love to use them and they can be very appealing. At minimum then, it is best to understand their limitations, and be cautious about their use.

3.10 Plotting Tables

Recall from the previous chapter that tables can be generated quite easily in R using the `table()` function. One can combine the `table()` function with that of `plot()` to generate a graph of the given table. For example, for the Galton data, we can obtain cases for each value on the `parent` variable:

```
> table.parent <- table(Galton$parent)
> table.parent
```

```
  64 64.5 65.5 66.5 67.5 68.5 69.5 70.5 71.5 72.5    73
  14   23   66   78  211  219  183   68   43   19     4
```

That is, there are 14 values of 64, 23 values of 64.5, 66 values of 65.5, and so on.

Using the `plot()` function on the object `table.parent`, we can generate nice visual summaries of the distribution:

```
> plot(table.parent)
> barplot(table.parent)
```

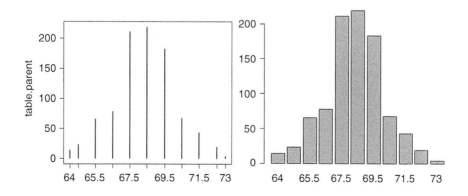

Exercises

1   Generate two vectors, and use the `plot()` function to obtain a basic plot. Then, using a `lines()` statement, add a line to the plot.

2   Use the `rnorm()` function to generate a random series of 100 digits, then use `some()` to view a few of these digits.

3   Generate a histogram of the 100 digits produced in #2, and add a rug plot to the histogram.

4   Again using `rnorm()`, generate a second vector, then obtain a scatterplot of this variable against the variable produced in #2.

5   Adjust the plotting character in the plot produced in #4, exploring a few options for `pch` as listed in the chapter.

6   In the chapter, a sunflower plot of parent height as a function of child height was generated. Produce a sunflower plot of child height as a function of parent height, and comment on the information the plot is conveying.

7   Plot parent height against child height using `qplot()`, and employ the geom "`jitter`." Communicate in general what the plot is depicting.

8   Generate a stem-and-leaf plot of the variable `parent`, then a boxplot of the same variable. Interpret both plots. Adjust the scale of the stem plot to explore options. What scale do you think is best to depict the data? Why?

9   On the iris data, conduct a Shapiro–Wilk test on the variable `setosa`. What is your decision regarding normality?

10   On the achievement data featured in the chapter, obtain a violin plot of `ac` (achievement scores). Compare the plot to a boxplot of the same variable. What information does the violin plot provide that the corresponding boxplot does not?

4

Means, Correlations, Counts

Drawing Inferences Using Easy-to-Implement Statistical Tests

LEARNING OBJECTIVES

- How to compute z scores and areas under the normal curve.
- How to plot normal distributions.
- How to compute and evaluate a Pearson product-moment correlation and a Spearman correlation.
- How to compute alternative correlations in R such as biserial and point-biserial.
- How to compute one-sample, two-sample, and paired-samples t-tests.
- How to use R to compute a binomial test and tests on contingency tables.
- How to plot data for categorical data using mosaic plots, radar charts, etc.
- How to use Cohen's Kappa for measuring agreement between raters.

In this chapter, we survey a variety of the more popular statistical tests for dealing with a variety of data situations you might find yourself in, and need a relatively simple way to analyze such data without having to invoke a full statistical model as in the analysis of variance or regression. Many of the later chapters develop a statistical model or methodology. This chapter is a bit different. Instead, this chapter simply features a handful of statistical procedures that can be immediately applied to a given data situation. For each test, we list the objectives of the test, the kind of data suitable for the test, and how to summarize findings. In essence then, this chapter is about providing a useful "toolkit" of essential procedures that you can apply fast, and get immediate results without knowing too much of the theory behind the procedure.

4.1 Computing z and Related Scores in R

Data analysts often speak of "standardizing" a distribution. But what exactly does this mean? Standardizing a distribution means performing a **linear transformation** on the given distribution such that the new distribution will have a mean

Univariate, Bivariate, and Multivariate Statistics Using R: Quantitative Tools for Data Analysis and Data Science, First Edition. Daniel J. Denis.

of 0 and a standard deviation of 1. It is a linear transformation because the actual relative distances between points before and after the transformation remains the same. These standardized scores are referred to as **z scores**. One reason they are useful is that they allow a relative comparison of where an individual score stands relative to other scores in a distribution. To compute a z score for a given observation, we subtract the original mean of the distribution we are working with, and divide by the standard deviation of the distribution. That is, the computation of a z score is carried out as follows:

$$z = \frac{x - \mu}{\sigma}.$$

We can see that if x is equal to the mean of the distribution, μ, then the numerator of the ratio is 0, and since 0 divided by σ is equal to 0, the resulting z is also equal to 0. So how do we generate z scores in R? We can do so using the `scale()` function. Recall the Galton data and the variable of parent height:

```
> head(parent)
[1] 70.5 68.5 65.5 64.5 64.0 67.5
```

The mean and standard deviation of `parent` are equal to

```
> mean(parent)
[1] 68.30819
> sd(parent)
[1] 1.787333
```

Suppose we wish to compute the z score for a first raw score, that of 67.5, the last observation in our brief list above. Our computation is:

$$z = \frac{x - \mu}{\sigma} = \frac{67.5 - 68.30819}{1.787333} = -0.45.$$

The obtained z score is negative, meaning that it lies **below the mean** z of 0. We can see that this makes good sense, since 67.5 is below the mean of 68.31. Let's get R to compute z scores for us, and verify that it did so correctly for the sixth observation in our data:

```
> scale(parent)
            [,1]
[1,]   1.2263019
[2,]   0.1073165
[3,]  -1.5711616
[4,]  -2.1306543
[5,]  -2.4104007
[6,]  -0.4521762
```

We can see that the sixth observation has a z score equal to −0.45, which agrees with the z score we generated manually. Let's now compare a distribution of raw scores for `parent` to one of standardized scores, side-by-side:

```
> hist(parent)
> hist(scale(parent))
```

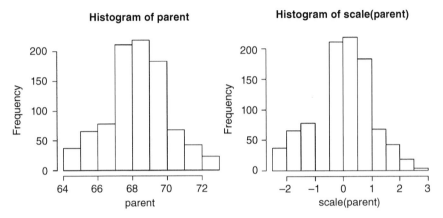

Histogram of parent   **Histogram of scale(parent)**

At first glance, it would appear the z transformation had the effect of changing the distribution somewhat. But this is an illusion! The distribution has not changed. In plotting the data, R simply **rescaled** the x-axis somewhat as to make it appear as if the distribution has changed. However, as mentioned, the computation of a z score is simply a linear transformation of the original data, and hence, the two distributions will have the same shape. One way of understanding what a linear transformation is, is to obtain a plot of z against `parent`:

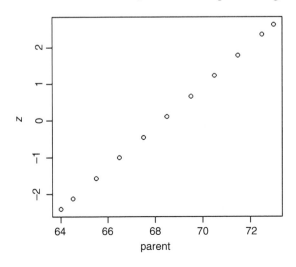

We can see that as `parent` increases, z also increases in a directly **linear** fashion. That is, the relative distances between points on the `parent` axis is the same as the relative distances between points on the z axis. The key point here is to recognize that z did not change the distribution; it merely transformed it into a new distribution with a new mean and standard deviation.

This is all fine, but how can we use z scores anyway? Though z scores are used throughout statistics in a variety of ways and in different settings and contexts, one way in which we can use them is for simple hypothesis testing in a sample of data points. For a given raw score, we can convert to a standardized score, and determine the area above and below the given score. Referring back to Galton's data on heights, recall the mean parent height was equal to 68.31, with a standard deviation equal to 1.79. Suppose at random, a height of 75.00 is drawn. We may ask the question:

What is the probability of observing a height of 75.00 or greater from a normal distribution having mean 68.31 and standard deviation 1.79?

We can determine this probability quite easily. The z score associated with a raw score of 75 is $(75 − 68.31)/1.79 = 3.74$. That is, in a normal distribution, the raw score of 75 is 3.74 standard deviations above the mean. Getting back to our original question, we can determine the probability of such a score in a standard normal distribution by computing the area under the curve above 3.74. Computing areas under curves goes by the name of **integration** in calculus, and most of these areas are readily available for the standardized normal distribution at the back of most introductory statistics textbooks. Fortunately, R can compute these proportions under the curve on the spot. We can use the function `pnorm()`, which is the **cumulative distribution function** (or "cdf" for short) for the normal distribution, to obtain the probability under the normal distribution. For instance, for $z = 0$, the density above this value is half of the normal curve, giving a value of 0.5:

```
> pnorm(0)
[1] 0.5
```

For our value of $z = 3.74$, we know from our knowledge of the normal distribution that very little area under the curve will lie above this value. By computing `pnorm(3.74)`, R will tell us how much area exists below this value:

```
> pnorm(3.74)
[1] 0.999908
```

Hence, the area above 3.74 is computed by

```
> 1-pnorm(3.74)
[1] 9.201013e-05
```

As we guessed it would be, this is an extremely small area. Hence, the probability of observing a raw score of 75.00 from this distribution is equal to 0.0000920. If a score had been drawn at random, and it equaled 75.00, we would say the probability that this score could have come from this distribution with mean 68.31 and standard deviation 1.79 is exceedingly low, clearly much lower than the often arbitrary cutoff value of 0.05 (i.e. 0.0000920 is much smaller than 0.05). Hence, in a hypothesis-testing framework, we would reject the null hypothesis that the score was drawn from this population, and conclude that it was likely drawn from a different population. This is one way in which z scores can be used to evaluate the tenability of null hypotheses.

4.2 Plotting Normal Distributions

We can request R to plot **normal distributions** as well. The following uses the function seq() to plot a normal distribution with length on the x-axis equal to 100. You can experiment with varying the plotting parameters to see its effect on the corresponding plot:

```
> x <- seq(-5, 5, length = 100)
> norm <- dnorm(x)
> plot(norm)
```

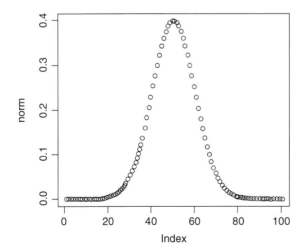

4.3   Correlation Coefficients in R

There are a number of correlation coefficients one can compute in R. It is imperative from the outset that you do not automatically associate the phrase **correlation coefficient** with that of **linear correlation**, as due to the variety of coefficients available, not all of them are meant to capture linearity. For example, as we will see, it is entirely possible for the correlation coefficient **Spearman's rho** to attain very high values, and yet the relationship between the two variables may not be linear. Hence, if you come across a statement in the research literature stating that the correlation between two variables is equal to 0.90, for instance, be sure to learn from that research paper **which correlation was computed**; otherwise, you may incorrectly and automatically assume the relationship is strongly linear, which it may not be. Having said that, the most common measure of correlation reported across the sciences is that of **Pearson**'s *r*, which we survey next and learn how to compute in R.

The **Pearson product-moment correlation**, or Pearson's *r* for short, is a measure of the **linear** relationship between two continuous variables. If your variables are categorical rather than continuous, then computing Pearson's *r* is typically incorrect, so be sure your two variables have sufficient distribution to be at least approximated by a continuous distribution before computing *r*.

The term "correlation" may refer to a wide variety of correlation coefficients; however, the most popular is by far, the Pearson product-moment correlation coefficient, or Pearson's r for short. However, be cautious about automatically associating the word "correlation" with Pearson's r as other correlation coefficients may report vastly different strengths of relationship. Pearson's r is most appropriate for data on which both variables are continuous, and you wish to measure the linear relationship between them.

Formally, the Pearson product-moment correlation is defined by:

$$r = \frac{\frac{\sum_{i=1}^{n}(x_i - \bar{x})(y_i - \bar{y})}{n-1}}{s_x \cdot s_y} = \frac{\text{cov}_{xy}}{s_x \cdot s_y}.$$

The numerator of this equation is what is known as the **covariance**, which is defined as a sum of cross-products divided by $n-1$. Covariances are **scale-dependent**, which means that a large covariance may result in part due to variabilities in x and y, and not only the cross-product term $(x_i - \bar{x})(y_i - \bar{y})$. What this means pragmatically is that it is entirely possible to get a reasonably sized covariance between two variables, yet the degree of linear relationship still need not be exceedingly large. To deal with this, we **standardize the covariance** by dividing it by the product of standard deviations, $s_x \cdot s_y$. By doing so, we make the resulting number, called Pearson's r, **scale-independent**. The standardization also has the effect of placing limits of -1 to $+1$ on this new measure of Pearson's r. An r of -1 means that as values of x increase, values of y decrease in a perfect linear fashion. Contrarily, an r of $+1$ means that as values of x increase, so do values of y in a perfect linear fashion. A Pearson's r equal to 0 indicates the absence of a linear relationship among the variables. Be very careful to note that $r = 0$ does not necessarily imply the absence of a relationship between x and y, but only of a **linear** one. It is entirely possible to obtain $r = 0$, yet there still exist a relationship that is something other than linear. **Always plot your data** to reveal the form of the relationship, and never rely on numerical coefficients alone to describe them.

The following are some examples of scatterplots and correlations often depicted in introductions to statistics:

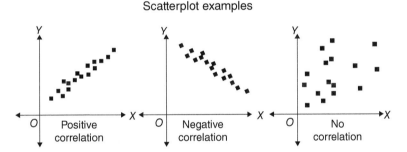

Scatterplot examples

It is also important to realize that though Pearson's r is computable on any sample of data, an inference of the corresponding population parameter is preferable (though not always required, see Yuan and Bentler, 2000) under the assumption of **bivariate normality**, which can often at least informally be verified by a visual inspection of the bivariate plot. More formal tests exist (see Johnson and Wichern, 2007), but are seldom used in practice. Pragmatically, unless there are obvious nonlinear trends in one's data, or visible outliers that have a high impact on the data, usually one can assume the data are at least **approximately** bivariately normal and proceed with the calculation and significance test of Pearson's r. The bottom line is that assumptions such as this are not truly ever satisfied perfectly, and if your data look reasonable, you're likely fine with drawing inferences.

To demonstrate its computation in R, recall once more Galton's plot of parent heights against child heights:

```
plot(parent, child)
```

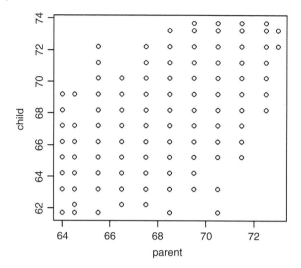

Computing Pearson's *r* in R (ha!) is very easy. Let's compute the correlation between parent height and child height in Galton's data:

```
> cor(parent, child)
[1] 0.4587624
```

By default, R computed a Pearson correlation. However, since there are numerous other correlation coefficients, to be most complete, we could have specifically designated that we wished a Pearson correlation:

```
> cor(parent, child, method = "pearson")
[1] 0.4587624
```

We will return shortly to considering alternatives to *r* and specify a different coefficient under method = "pearson". But first, what does this coefficient of 0.45 mean? It means that scores on parent tend to vary with scores on child. That is, we can conclude, at least for this particular scatterplot, that **on average, for a score above the mean on parent, it is more likely than not that a score for child is also above the average for child. Likewise, for a score below the mean of parent, we are more likely than not to have a score below the mean for child**. Notice however that a correlation of 0.45 is only about halfway to the maximum value of *r*, which is 1.0. Hence, there is a lot of **unexplained variation** going on as well. Sometimes researchers get all enthused about correlations that are rather small, yet fail to realize that most of what they are seeing in the empirical pattern is actually unexplained variation. For the correlation of 0.45, this definitely does not mean that if a parent is tall, the corresponding child **will** be tall. For one, look at all the observations in the lower right of the plot, where horizontal and vertical lines have been drawn in at the approximate means for parent and child, respectively:

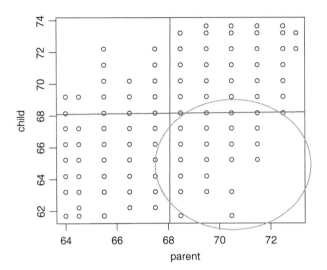

Notice that for a taller-than-average parent, there are plenty of adult children who end up less than the average height for adult children. Hence, the conclusion that if you have tall parents you will **necessarily** have tall children is definitely false. However, what we can say from this particular plot is that if you have relatively tall parents, you're in a sense more **likely** than not to have taller-than-not children. Such a conclusion is much weaker, but it is by far more accurate and ethical. Be wary of researchers who like to conclude "too much" from their data to support their theory of the day! Science isn't (read: shouldn't be) about establishing a brand, it's about discovering empirical and theoretical realities. See Kuhn (2012) for a thorough discussion of how science often isn't quite as "objective" as we sometimes think it is.

In addition, it must be emphasized that you can never conclude any kind of **causal statement** from a correlation coefficient alone. If the research situation merits an inclination toward causal language, then that's different, but you can never deduce **ontological causation** from a mere correlation coefficient (or any other statistic, for that matter). For Galton's data, because height is largely genetically transmitted, it wouldn't be that terrible to conclude that parent height is a **causal force** in determining child height. However, that conclusion is not being made as a result of the correlation coefficient. Rather, it is being made because we know substantively that attributes such as height are transmitted **biologically**, and thus the **deduction of causation** is based on the research variables and substantive setting rather than because the two variables were found to be correlated. Causation is a very complex topic, and you should never use it lightly in a research setting. Philosophers who study these things in depth (more depth than we probably ever care to know!) cannot even agree on what causation even **is**, exactly, so if you use the word "causation" in your research, be prepared to unpack (and defend) it for your audience (hint: it will not be easy, and hence you are better off not talking about causation at all in your presentation or publication).

Causation can never be inferred from a correlation coefficient alone. A correlation, no matter how large, does not necessarily imply causation. However, if one can deduce causation from other means, for instance substantive knowledge of the research situation or research variables, then causation becomes a more plausible argument. However, keep in mind that there are numerous theories of causation, and philosophers (most of whom are very deep thinkers!) have yet to agree on what causation actually "means" in a unified definition, so if you are to use the word at all, at minimum, you should have a strong defense for what you mean by the term.

4.4 Evaluating Pearson's *r* for Statistical Significance

Above, we obtained the correlation of 0.45 between parent and child height. However, this is only the correlation in the **sample** of observations. Suppose now we would like to evaluate the null hypothesis that the population correlation coefficient is equal to 0. We can do this using the function `cor.test()`:

```
> cor.test(parent, child, method = "pearson")

Pearson's product-moment correlation

data:  parent and child
t = 15.711, df = 926, p-value < 2.2e-16
alternative hypothesis: true correlation is not equal to 0
95 percent confidence interval:
 0.4064067 0.5081153
sample estimates:
      cor
0.4587624
```

Recall what `p-value < 2.2e-16` means: it is the probability of observing a sample correlation such that we have observed (of 0.4587624) under the assumption of the null hypothesis that the true correlation in the population is equal to 0. Since that probability is very low, certainly much lower than most conventional levels used in statistical tests (e.g. 0.05, 0.01, 0.001), we reject the null hypothesis and conclude the alternative hypothesis that the true correlation is not equal to 0 from which these sample data were drawn. We can also interpret the 95% confidence interval as the range of plausible values for the parameter, likely to lie between the lower limit of 0.406 and the upper limit of 0.508.

4.5 Spearman's Rho: A Nonparametric Alternative to Pearson

We have seen that Pearson's *r* is effective at capturing the linear association or relationship between two continuous variables. Many times, however, relationships are **nonlinear** in form. Though there is nothing preventing us from computing Pearson's correlations on nonlinear relationships, such correlations would not "capture" the relationship or adequately represent the pattern of how the variables vary together. We need a different measure of correlation to measure a potentially nonlinear pattern.

One such type of correlation coefficient is **Spearman's rho**. Spearman's correlation coefficient is suitable for situations where the relationship between *y* and *x*, while possibly not being linear, is nonetheless what we will refer to as **monotonically increasing or decreasing**, which means as *x* increases (or decreases), *y* also increases (or decreases), but not necessarily in a linear pattern. The "monotonicity" here refers to the increase (or decrease) in one variable with an increase (or decrease) in the other, though it does not guarantee **linear monotonicity**. That is, a monotonic increase or decrease does not imply such monotonicity has a linear shape.

A simple example will help demonstrate the differences between a Spearman and a Pearson coefficient. Consider data on movie preference rankings taken from Denis (2016). In Table 4.1, movie favorability is ranked from highest (1) to lowest (5) for both Bill and Mary. For example, while Bill likes Star Wars best, Mary's favorite movie is Scarface. In brackets are actual favorability scores measured on each individual for each movie, scored from 1 to 10. Bill scored a 10 for Star Wars (maximum favorability), while Mary scored a 9.7 for Scarface. Hence, in the first case, we have **rankings**, whereas in the second case are actual favorability **scores** measured on a continuous scale.

Table 4.1 Favorability of movies for two individuals in terms of ranks.

Movie	Bill	Mary
Batman	5 (2.1)	5 (7.6)
Star Wars	1 (10.0)	3 (9.0)
Scarface	3 (8.4)	1 (9.7)
Back to the Future	4 (7.6)	4 (8.5)
Halloween	2 (9.5)	2 (9.6)

Actual scores on the favorability measure are in parentheses.

Let's first correlate favorability scores between the two raters, computing both a Pearson and a Spearman. First, we generate the data and compute a Pearson:

```
> bill.fav <- c(10.0, 9.5, 8.4, 7.6, 2.1)
> mary.fav <- c(9.7, 9.6, 9.0, 8.5, 7.6)
> cor(bill.fav, mary.fav, method = "pearson")
[1] 0.9551578
```

The Pearson correlation is equal to 0.955. A plot of this relationship matches about what we would expect for such a correlation: `> plot(bill.fav, mary.fav)`

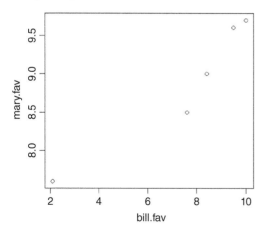

We can see that the relationship is not perfectly linear, as evidenced by the fact that Pearson's *r* is not a perfect +1.0. Indeed, the relationship appears to be somewhat nonlinear in form, though we would agree that as Bill's scores increase at each turn, so do the scores of Mary. In other words, there is a **monotonically increasing relationship between scores for Mary and those for Bill**, even though the relationship is not perfectly linear. If Spearman's rho captures what we advertised it to capture, then this monotonic relationship should be reflected in a perfect correlation coefficient of +1.0. Let's compute Spearman to confirm this:

```
> cor(bill.fav, mary.fav, method = "spearman")
[1] 1
```

As we can see, Spearman's results in a perfect correlation. This is so because it is capturing the monotonically increasing relationship between variables. Another interpretation of Spearman here is that it is the **Pearson correlation on ranked data**. Because Spearman is ignoring the magnitude differences between the scores, but rather focusing only on their ordinal position (i.e. their "rank"), if we computed a Pearson correlation on the ranks, we would obtain the same as if we computed a Spearman.

4.6 Alternative Correlation Coefficients in R

In addition to Pearson and Spearman, there are a whole host of alternative correlation coefficients that can be computed in R. For instance, a **point-biserial correlation** coefficient can be used when one of the variables is **dichotomous** and the other **continuous**. Some variables, of course, are naturally dichotomous, such as survival (yes versus no), while others can be made dichotomous by re-operationalizing them from continuous to dichotomous. For instance, suppose we classified those above 6 feet in height to be "tall" and those below it to be "short." Such an example would be considered an artificial dichotomy instead of a natural one, and when correlating such a variable with a continuous variable we would refer to the resulting coefficient as simply a **biserial correlation** instead of a point-biserial one as was true for when we had a naturally occurring dichotomy on both variables. For details on biserial correlations, and many other correlation coefficients, see Warner (2013).

As an example of a point-biserial correlation in R, consider the following data on grades (0 = failure, 1 = pass), and amount of time spent studying for a quiz, in minutes:

```
> grade <- c(0, 0, 0, 0, 0, 1, 1, 1, 1, 1)
> study.time <- c(30, 25, 59, 42, 31, 140, 90, 95, 170, 120)
> grade.data <- data.frame(grade, study.time)
> grade.data
```

```
      grade study.time
1       0          30
2       0          25
3       0          59
4       0          42
5       0          31
6       1         140
7       1          90
8       1          95
9       1         170
10      1         120
```

The point-biserial correlation can be computed by simply computing the ordinary Pearson correlation, and naming it **point-biserial**:

```
> cor.test(grade, study.time)

        Pearson's product-moment correlation

data:  grade and study.time
t = 5.3515, df = 8, p-value = 0.0006846
alternative hypothesis: true correlation is not equal to 0
95 percent confidence interval:
 0.5740098 0.9724262
sample estimates:
      cor
0.8841088
```

Hence, the point-biserial correlation between grade and study.time is 0.88. The p-value associated with the correlation is 0.0006846, and hence we can reject the null hypothesis that the true correlation in the population is equal to 0.

4.7 Tests of Mean Differences

4.7.1 t-Tests for One Sample

In a t-test for a single sample, we are interested in evaluating the probability that the observed data could have reasonably been drawn from a population with a particular specified mean. A t-test is most useful when we do not have knowledge of the population standard deviation, and especially when we are working with smaller-than-not sample sizes. To demonstrate a simple *t*-test for one sample, consider the following small sample data on IQ scores:

```
> iq
[1] 105   98 110 105   95

> t.test(iq, mu = 100)
```

In the t-test() function, we first specify the vector containing the data, which is iq, then mu = 100 indicates the value of the population mean we wish to test.

```
        One Sample t-test

data:  iq
t = 0.96495, df = 4, p-value = 0.3892
alternative hypothesis: true mean is not equal to 100
95 percent confidence interval:
  95.11904 110.08096
sample estimates:
mean of x
    102.6
```

We note the following conclusions based on the output:

- We see that the obtained *t* statistic is equal to 0.96495, evaluated on 4 degrees of freedom, yielding a *p*-value of 0.3892.
- The null hypothesis was that the sample was drawn from a population with mean equal to 100, and the alternative hypothesis was that the true mean is not equal to 100. Since $p = 0.3892$ is relatively large (certainly not smaller than some conventional level such as 0.05 or 0.01), we fail to reject the null hypothesis, and therefore do not have evidence that the true mean is not equal to 100.
- The 95 percent confidence interval with limits 95.11904 and 110.08096 implies we are 95% confident that the true population mean lies within these values. That is, the range 95.12 and 110.08 is plausible for the true mean. We note that since the population mean of 100 lies within this range, this yet again supports the idea that we do not have evidence to reject the null hypothesis.
- Finally, R reports the mean of the sample, equal to 102.6.

4.7.2   Two-Sample *t*-Test

A two-sample test is used to evaluate the null hypothesis that two population means are equal, or equivalently, that both samples were selected from the same population. The alternative hypothesis is that the samples were drawn from two different populations. As an example of a two-sample *t*-test, consider the following data and associated output, where grade.0 represents grades from one class, and grade.1 represents grades from a different class, **independent** of the first class. That is, for this *t*-test, we are assuming the samples are not related or paired in any way (this is in contrast to the paired-samples *t*-test we will perform shortly where data are sampled in pairs and not independently):

```
> grade.0 <- c(30, 25, 59, 42, 31)
> grade.1 <- c(140, 90, 95, 170, 120)
> t.test(grade.0, grade.1)

        Welch Two Sample t-test

data:  grade.0 and grade.1
t = -5.3515, df = 5.3094, p-value = 0.002549
alternative hypothesis: true difference in means is not
equal to 0 95 percent confidence interval:
 -126.00773   -45.19227
sample estimates:
mean of x mean of y
     37.4      123.0
```

We summarize the output:

- The obtained statistic is equal to −5.3515, evaluated on degrees of freedom equal to 5.3094. Notice that by default, R conducts the **Welch test**, which is an adjustment for **inequality of variances**. In addition to assuming the data in each population arise from approximately normal distributions, the *t*-test also assumes that the variances in each population are more or less equal (more on this assumption when we discuss ANOVA in Chapter 6). In conducting the *t*-test as we did above, R automatically assumes we have violated the assumption (we will test it in a moment to check). Returning to the difference in means, the obtained *p*-value is equal to 0.002549, and hence we have evidence to reject the null hypothesis at 0.05 and infer the alternative hypothesis that the true difference in means in the population is not equal to 0.
- The 95% confidence interval has limits of −126.00773 and −45.19227, and hence we are 95% confident that the true mean difference lies between these limits.
- Finally, R provides us with the sample estimates of the means per group. We can compute the mean difference from these, as 37.4 − 123.0 = −85.6. Notice that this number of −85.6 lies at the center of the confidence interval.

Cohen's d for this data can be computed using the **lsr** package:

```
> library(lsr)
> cohensD(grade.0, grade.1)
[1] 3.384563
```

We conclude that the mean difference in standard deviation units is 3.38. Just how big of an effect is this? Recall that how big an effect is typically depends on the field you're working in; however, under most circumstances, the above would be considered a very large effect size. Recall that Cohen himself generally considered values of **d** equal to 0.8 or more to be large effects if one absolutely had to adopt some arbitrary convention for determining size. I do not recommend you allow arbitrary guidelines to guide your work. If you don't know what a large effect size is in your area of investigation, you need to familiarize yourself with your science a bit more.

4.7.3 Was the Welch Test Necessary?

To verify whether doing a Welch test was necessary, we can easily verify the assumption of variances using the `var.test` function in R:

```
> var.test(grade.0, grade.1)

    F test to compare two variances

data:  grade.0 and grade.1
F = 0.16831, num df = 4, denom df = 4, p-value = 0.1126
alternative hypothesis: true ratio of variances is not
equal to 1 95 percent confidence interval:
 0.01752408 1.61654325
sample estimates:
ratio of variances
          0.1683105
```

The function `var.test()` evaluates the assumption that the variances are equal. Since the *p*-value of 0.1126 for the test is larger than a conventional level such as 0.05, we fail to reject the null hypothesis of equal variances, and thus have no reason to doubt the null. We could have then, had we wanted, conducted the prior *t*-test assuming variances are equal and not implemented the Welch. To do this, we could have simply told R that we assumed variances were equal:

```
> t.test(grade.0, grade.1, var.equal = TRUE)
```

```
Two Sample t-test

data:  grade.0 and grade.1
t = -5.3515, df = 8, p-value = 0.0006846
alternative hypothesis: true difference in means is not
equal to 0 95 percent confidence interval:
 -122.48598  -48.71402
sample estimates:
mean of x mean of y
    37.4     123.0
```

We note now that the test was performed on 8 degrees of freedom, yielding a *p*-value of 0.0006846, much smaller than the previous Welch-corrected *p*-value. The degrees of freedom are larger in this case because the test doesn't "punish" us (in the form of the Welch correction in this case) for a potential violation of variances. The Welch test essentially lowered the degrees of freedom making it a harder test to pass, which explains the higher *p*-value obtained.

4.7.4   *t*-Test via Linear Model Set-up

We could have also performed the *t*-test in a more **linear model** framework (see Chapter 7), that is, instead of each group having its own vector of data points, we could have put all the data into a single vector and the grouping variable into its own vector, like this:

```
> studytime <- c(30, 25, 59, 42, 31, 140, 90, 95, 170, 120)
> grade <- c(0, 0, 0, 0, 0, 1, 1, 1, 1, 1)
> t.test(studytime ~ grade)
```

If you print the output of the test, you will see it matches our previous output for which we used a Welch correction. As we did earlier, to turn the correction off, we simply adjust by:

```
> t.test(studytime ~ grade, var.equal = TRUE)
```

4.7.5   Paired-Samples *t*-Test

The **paired-samples *t*-test** is used when we have paired data and wish to evaluate a null hypothesis that the mean difference is equal to 0. The requirement for this test is that data are **sampled in pairs**, meaning that the data in the first column is **associated** or **related** in some sense **naturally** with the data in the second column. This is referred to as **blocking** or **nesting**, also relevant in repeated-measures models, which will be the topic of Chapter 6 on the analysis of variance. In these models, there is maximal relatedness between columns since they are paired by the same subject. For the following data, **time** is the response variable, and such observations are nested within **rat** in a learning task. Each rat undergoes three trials, and hence for each rat, we have a total of three measurements (one measurement per trial):

```
> learn <- read.table("learning.txt", header = T)
> learn
   rat trial time
1    1     1 10.0
2    1     2  8.2
3    1     3  5.3
4    2     1 12.1
5    2     2 11.2
6    2     3  9.1
7    3     1  9.2
8    3     2  8.1
9    3     3  4.6
10   4     1 11.6
11   4     2 10.5
12   4     3  8.1
13   5     1  8.3
14   5     2  7.6
15   5     3  5.5
16   6     1 10.5
17   6     2  9.5
18   6     3  8.1
```

We will analyze these data further in Chapter 6 and conduct a full repeated measures on them. For now, we are only interested in learning whether there might be mean differences across trials 1 and 2:

```
> trial.1 <- c(10, 12.1, 9.2, 11.6, 8.3, 10.5)
> trial.2 <- c(8.2, 11.2, 8.1, 10.5, 7.6, 9.5)
> t.test(trial.1, trial.2, paired = TRUE)

Paired t-test

data:  trial.1 and trial.2
t = 7.2012, df = 5, p-value = 0.0008044
alternative hypothesis: true difference in means is not
equal to 0
```

```
95 percent confidence interval:
 0.7073371 1.4926629
sample estimates:
mean of the differences
                    1.1
```

Since the obtained *p*-value is very small, we reject the null hypothesis of equality between means, and conclude that the true difference in means is not equal to 0.

T-tests for one sample are used to evaluate the null that a sample has a given population mean. There is only one sample collected. A two-sample t-test is used to evaluate the null that two independent samples were drawn from the same population. There are two samples collected. A paired-samples t-test is used when data are sampled in pairs, which implies that knowing one observation of the pair reveals information about the second. Paired samples is an example of blocking or nesting, and is prevalent in repeated-measures models, to be surveyed in Chapter 6.

4.8 Categorical Data

Up to now, most of our data has been of the form where we had to **measure** the data, such as values on IQ or depression, etc. That is, some mechanism and rules existed such that numbers were assignable to values of a given variable (this is what measurement is). However, sometimes data come in the form of **counts** across a variety of categories, usually mutually exclusive. For instance, we might ask the question of how many people in our sample are male. How many female? Are the counts of males and females proportionally different from state to state across America? These types of questions feature **categorical data** and require different analytical methods than those typically used on measurement data, which is often continuous (or pseudo-continuous) in nature.

4.8.1 Binomial Test

Binomial tests can be considered special cases of the more elaborate **chi-squared goodness of fit test**. In a full chi-squared test, we typically have several mutually exclusive categories. In a binomial test, we always have only two. A classic example of where a binomial test may be useful is with coin flips. You may recall we referred to coin-flipping examples at the outset of this book, and asked the question of whether or not the coin we experimented with was fair. Imagine we start off by assuming a coin is fair, that is, the probability of heads is equal to the probability of tails, both equal to 0.5. We then flip the coin 5 times and obtain 2 heads. We ask the following inferential question:

What is the probability of getting 2 heads out of 5 flips on a fair coin?

Under the assumption of a fair coin, we would probably expect somewhere between 2 and 3 heads on 5 flips of a fair coin. That is, if the coin is indeed fair in reality, then we should expect approximately equal heads and tails. But we obtained 2 heads. What is the probability of this result? We can easily evaluate these data using the binom.test() function, where we first enter the number of "successes," then the number of total trials, then the probability of a success on any given trial:

```
> binom.test(2, 5, p = 0.5)
```

The "2" above is the number of successes (for our data, heads), and the "5" is the number of total trials. The probability of a success on any given trial is set at 0.5, that is $p = 0.5$. Note that this is not the traditional *p*-value associated with the significance test such as 0.05 or 0.01. The $p = 0.5$ here refers to the probability of heads under the null hypothesis. That is, the probability of success under the assumption that the coin is fair. The results of the binomial test follow:

```
Exact binomial test

data:   2 and 5
number of successes = 2, number of trials = 5, p-value = 1
alternative hypothesis: true probability of success is not
equal to 0.5
95 percent confidence interval:
 0.05274495 0.85336720
sample estimates:
probability of success
              0.4
```

We interpret the above output:

- For 2 successes out of 5 trials in which the probability of a success on any given trial is equal to 0.5, the obtained *p*-value is equal to 1 (with point probability equal to 0.3125), and hence we have no evidence to reject the null hypothesis. That is, we do not have evidence to suggest the probability of success is not equal to 0.5.

Note carefully how the situation would change if we adjusted the probability of a success from $p = 0.5$ to something much different, such as $p = 0.9$. That is, instead of the assumed probability of heads on the coin being equal to 0.5, suppose we set it at 0.9 (i.e. such that we would expect approximately 9 heads out of 10 flips). For this situation, the test reveals:

```
> binom.test(2, 5, p = 0.9)
Exact binomial test

data:  2 and 5
number of successes = 2, number of trials = 5, p-value =
0.00856
alternative hypothesis: true probability of success is not
equal to 0.9
95 percent confidence interval:
 0.05274495 0.85336720
sample estimates:
probability of success
              0.4
```

Notice in this case, the obtained p-value is equal to 0.00856, which is statistically significant. Why is it statistically significant now and not before when p was set at 0.5? Because expectation under the null set at 0.90, a proportion of 0.4 (i.e. 2 heads out of 5 flips) is a much more **unlikely** event. The result of 2 heads out of 5 flips was **likely** because the null was $p = 0.5$. If we change the null, however, the probability of the data (the number of heads, in this case) also changes. This example serves to emphasize that whether or not data are likely or unlikely depends entirely not only on the obtained data, but also on **how the null hypothesis is defined**. We will revisit the binomial test when we survey the **sign test** in Chapter 13. The sign test is a nonparametric test that features the use of the binomial distribution.

The simple example of a binomial test on the probability of heads on a coin is a powerful demonstration of how the probability of the data is determined based not only on the obtained data, but on what null hypothesis is put forth. For example, 2 heads out of 5 flips is a very common event under a null hypothesis that the probability of heads is equal to 0.5, but a very unlikely event under the null hypothesis that the probability of heads is equal to 0.9. The take-away lesson is that one should never interpret a p-value without first knowing what the null hypothesis was on which that p-value was computed.

4.8.2 Categorical Data Having More Than Two Possibilities

We have just featured the binomial test as a simple test to evaluate whether counts were approximately equally distributed across two mutually exclusive categories. Our two categories were "head" and "tail" corresponding to sides of a coin. Most often, however, categorical data present themselves in **contingency tables**, where entries in the table represent the counts for each joint category on each variable.

For example, consider the following 2×2 contingency table:

	Condition present (1)	Condition absent (0)	
Exposure yes (1)	20	10	30
Exposure no (2)	5	15	20
	25	25	50

These fictitious data come from Denis (2016, p. 92) and are described as follows:

- The column variable is that of **condition** (such as a disease), which can be present (1) or absent (0).
- The row variable is that of **exposure** (e.g. to a toxin), of which the possible values are "yes" and "no."
- The numbers in the cells are **counts** corresponding to each cell combination. For example, there are 20 cases that have been exposed and have the condition present, while there are 10 cases who have been exposed who do not have the condition present. The total number exposed is located in the margin, equal to $20 + 10 = 30$.
- A total of 20 cases have not been exposed ($5 + 15 = 20$), and an equal number of cases (25) have the condition present versus absent.
- The total number of cases in the data is equal to 50.

In the following, we build the contingency table in R, by first generating the corresponding matrix, and then requesting a chi-squared test using the `chisq.test()` function to evaluate whether there is an **association** between the condition variable and the exposure variable:

```
> diag.table <- matrix(c(20, 5, 10, 15), nrow = 2)
> diag.table

     [,1] [,2]
[1,]   20   10
[2,]    5   15
```

We see that the above matrix corresponds to the 2×2 contingency table above. We now request the corresponding test:

```
> chisq.test(diag.table, correct = F)

        Pearson's Chi-squared test

data:  diag.table
X-squared = 8.3333, df = 1, p-value = 0.003892
```

The obtained *p*-value is 0.003892, and is thus statistically significant. R does not provide us with the critical value for the test, nor do we need to know it since we can simply observe the obtained *p*-value. However, if we wanted to know what the critical value is for a single degree of freedom in this case, we could check the appendix of most introductory statistics books, or we could simply ask R to tell us what it is by computing qchisq(0.95, 1), where 0.95 is the confidence level, which implies that the type I error rate is the classic 0.05, and 1 are the degrees of freedom for the test. The function qchisq() computes quantiles of the chi-squared distribution. When you try this in R, you'll see the critical value comes out to be 3.841459. It's not surprising then that the *p*-value is so low, since the value of 8.33 clearly exceeds 3.84.

In specifying correct = F, we negated what is known as **Yates' correction for continuity**, which essentially helps to make probabilities for discrete events a bit more continuous, and is sometimes preferred. Had we included the correction, it would have raised the *p*-value from 0.003 to 0.009 (try it for yourself by replacing correct = F with correct = T). When expected cell counts (i.e. those counts we would expect per cell under the null hypothesis of no association) are exceedingly small (e.g. less than 5 expected in a given cell), one can also perform **Fisher's exact test** using fisher.test(diag.table) for our data.

As a measure of association between condition and exposure, we can compute a **phi coefficient**:

```
> library(psych)
> phi(diag.table, digits = 3)
[1] 0.408
```

The value of 0.408 is interpreted as a Pearson correlation coefficient when both variables are binary. The phi coefficient could have also been obtained via the **vcd** package, along with a few other related statistics useful for measuring association in contingency tables:

```
> library(vcd)
> assocstats(diag.table)
                    X^2 df   P(> X^2)
```

```
Likelihood Ratio 8.6305   1 0.0033059
Pearson          8.3333   1 0.0038924

Phi-Coefficient    : 0.408
Contingency Coeff.: 0.378
Cramer's V         : 0.408
```

We saw above that the *p*-value was quite low, but as always, to get a sense of the effect, it helps to plot the data. We can visualize findings in R using the **vcd** package, and requesting what is known as a **mosaic plot**:

```
> library(vcd)
> mosaic(diag.table)
```

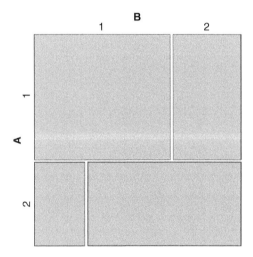

The mosaic plot displays **areas** corresponding to cell frequencies. That is, the area of the cells is proportional to the given cell total. For example, the cell in row 1, column 1 has a large area because it represents a total of 20 cases. The cell next to it in row 1, column 2 has a smaller area since it only represents 10 cases. Notice that the areas in the first row also correspond proportionally to the total in the margin of 30. That is, the cell with 20 is approximately 66% of the first row total, which corresponds to the area we see in the plot. Likewise, the second row areas correspond to frequencies of 5 and 15, where 5 is 25% of the marginal total of 20.

Another very useful way of portraying information in 2 × 2 tables is via a **fourfold plot**, also from the vcd library:

```
> library(vcd)
> fourfold(diag.table)
```

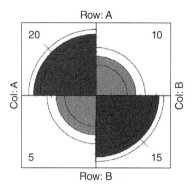

We can see that the frequencies in each quadrant correspond to the frequencies in each cell of the table. For more information on this plot, including how to interpret odds ratios and confidence intervals from it, as well as further options for plotting categorical data in R, see Michael Friendly's (excellent) website www.datavis.ca.

4.9 Radar Charts

So-called **radar charts** are another alternative for plotting categorical data. To demonstrate, we will create some hypothetical data representing grades for a statistics class, where the top grade possibility is A+ and the minimum grade a student can receive is F. Adapting code from Holtz (2018 – *The R Graph Gallery*) we load the package **fmsb**, and then proceed to build a sample data frame:

```
library(fmsb)
data = data.frame(matrix(sample(1:100, 10, replace = T),
ncol = 10))
colnames (data) = c("A+", "A", "A-", "B+", "B", "B-",
"C+", "C", "C-", "F")
data = rbind(rep(100, 10), rep(0, 10), data)
data
```

	A+	A	A-	B+	B	B-	C+	C	C-	F
1	100	100	100	100	100	100	100	100	100	100
2	0	0	0	0	0	0	0	0	0	0
3	55	70	90	5	76	19	2	61	87	95

From the data frame, we see that 55 students received grades of A+, 70 grades of A, and so on. We obtain a basic radar chart (Holtz, 2018) as follows:

```
radarchart(data, axistype=1,
cglcol="grey", cglty=1, axislabcol="grey",
caxislabels=seq(0,100, 25), cglwd=0.8)
```

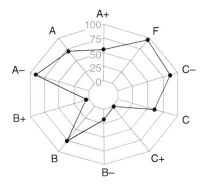

We can see from the above plot that the corresponding frequencies are represented. A radar chart such as this can be a powerful way to get an immediate sense of a distribution of counts. However, for most cases, the aforementioned bar charts and such will do just as good of a job and may be less ambiguous to interpret. Like pie graphs, the radar chart suffers a bit from the problem of not easily being able to decipher the magnitudes. For instance, if we only had the above chart to go by, it would prove difficult at first glance to know whether there were more grades of F versus A – or vice versa. Histograms generally do not have this problem, as it is usually much easier to detect which bin is higher. That is, the horizon is more apparent. Scientific research is about communicating results as clearly and as objectively as possible, not about who can produce the prettiest picture. Always be cautious about potentially "over-engineering" a graphic. Just because you can do it, doesn't mean it should be done, analogous to just because you can put 800 horsepower in a legal muscle car doesn't mean you should! (i.e. over-engineering can get you into trouble). More is not always better.

4.10 Cohen's Kappa

Cohen's kappa is a measure of **inter-rater agreement**, which means it measures the extent to which raters on some classification task agree in each other's ratings. As an example, imagine the case where there are two clinical psychology intern students who are attempting to classify the symptoms of disorders as having either a **psychological** or **biological** etiology. Suppose the frequencies of their ratings turn out as in the following table:

		Intern A		
		Psychological (1)	Biological (2)	Other (3)
Intern B	Psychological (1)	20	5	3
	Biological (2)	7	8	4
	Other (3)	7	3	5

To make sure you are reading the table correctly:

- We can see that both interns rated 20 cases as "psychological," both rated 8 cases as "biological," and both agreed on 5 other cases as "other." All of these numbers are represented along the main diagonal of the table, reading from top left to bottom right.
- In the off-diagonal, we see that Intern A rated 7 cases as psychological while Intern B rated those cases as biological. These frequencies are given in column 1, row 2.
- Intern A rated a total of $20 + 7 + 7 = 34$ cases as psychological, whereas Intern B rated a total of $20 + 5 + 3 = 28$ cases as psychological.

As we did earlier for the chi-squared data, to set the data up for Cohen's kappa in R, we first need to build the contingency table. We can easily use the `matrix()` function again to do this:

```
> intern.data <- matrix(c(20, 7, 7, 5, 8, 3, 3, 4, 5), 3, 3)
> intern.data
      [,1] [,2] [,3]
[1,]   20    5    3
[2,]    7    8    4
[3,]    7    3    5
```

We could have also specified the 3×3 matrix by specifically designating 3 rows and 3 columns (i.e. 3, 3, at the tail end of the statement):

```
intern.data <- matrix(c(20, 7, 7, 5, 8, 3, 3, 4, 5), nrow=3,
ncol=3)
```

In our earlier statement, where nrow = 3, R had to basically figure out on its own that there must be 3 columns instead of us telling it explicitly as we just did with 3, 3.

Having prepared the matrix for analysis, we can use the function `Kappa.test()` in the **fmsb** package to obtain the measure of agreement:

```
> library(vcd)
> Kappa.test(intern.data)
$`Result`

        Estimate Cohen's kappa statistics and test the null
        hypothesis that the extent of agreement is same as
        random (kappa=0)

data:  intern.data
Z = 2.583, p-value = 0.004898
```

```
95 percent confidence interval:
 0.05505846 0.45158606
sample estimates:
[1] 0.2533223

$Judgement
[1] "Fair agreement"
```

We summarize the results of the test:

- The null hypothesis is that the extent of agreement among the two raters is the same as one would expect by chance (i.e. "same as random" as indicated in the output).
- Kappa's test yields a value of 0.2533, and the associated *p*-value is 0.004898.
- Hence, we can reject the null hypothesis and infer that the degree of agreement between the two raters is beyond what we would expect by chance alone.
- As reported by R, the extent of agreement is classified as "fair agreement." However, as always, what determines whether an effect is large or small should depend on your opinion of it as a researcher. Perhaps other interns perform much better and as such, the obtained result is actually quite minimal relative to them? Remember that effect sizes are never interpretable "in a vacuum." A large effect in one field of study may actually be quite small in another.

Exercises

1 Discuss the nature of a *z*-score. What is it? What do *z*-scores accomplish, and why do they appear virtually everywhere in statistics and research?

2 Generate a series of 10 random numbers in R, and convert them to *z*-scores using the scale() function. Plot each distribution in a histogram, and compare them. What do you notice?

3 For a *z*-score of 1.5, what is the probability of obtaining a score greater than this in a normal distribution? Use R's pnorm() function to obtain your answer.

4 Discuss the nature of a Pearson correlation. Can you describe to someone what it measures? (Go beyond just saying "linear relationship" – be more specific and detailed.)

5 Discuss why standardizing the covariance makes an otherwise scale-dependent measure now dimensionless, and why this is important.

6 Your colleague says to you, "A Pearson correlation equal to 0 implies no relationship among variables." Do you agree? Why or why not? Explain.

7 Discuss the differences between Pearson and Spearman coefficients, and demonstrate on a sample of 10 random data points that Spearman is equivalent to Pearson's on ranked data.

8 Consider once more the data on `parent` in Galton's data. Conduct a one-sample *t*-test in R with a null hypothesis that the population mean is equal to 80. What is your conclusion?

9 Distinguish when to use a paired-samples t-test over an independent-samples *t*-test.

10 Generate a 2×2 contingency table with frequencies within each cell. Generate the frequencies such that the *p*-value for the resulting chi-squared test is a maximum value, then conduct the test in R to confirm.

5

Power Analysis and Sample Size Estimation Using R

LEARNING OBJECTIVES

- Understand the nature of, and determinants of statistical power.
- Learn how to estimate power and sample size for t-tests, ANOVA, and correlations as examples of power computations.
- How to translate effect size measures from R^2 to f^2 to enter relevant parameters in R to estimate power and sample size for an experiment or research study.
- Revisit significance testing in light of your knowledge of statistical power.

In this chapter, we will demonstrate some of the facilities in R for computing power and estimating required sample size for a few statistical models, focusing much of our attention on interpreting what the given power estimate and analysis means. This is so when you encounter or perform additional power analyses in your research for other designs, you will be well-equipped to understand the **principles at play**. In this chapter then, there is no data to analyze, as power analyses are conducted before the experiment or study begins. That is, you always want to estimate power and adequate sample size before you recruit your participants, so you can effectively plan a good study. An experiment deficient in power from the start will have little chance at rejecting a false null hypothesis.

However, before we begin, a discussion of what power is all about is in order, along with its determinants.

5.1 What Is Statistical Power?

Statistical power is the probability of rejecting a null hypothesis given that the null hypothesis is actually false. That is, if the null hypothesis under test is, in reality not true, **power is your probability of detecting that falsity**. Power is important especially if you place a lot of stock into p-values. Recall from Chapter 1 our

Univariate, Bivariate, and Multivariate Statistics Using R: Quantitative Tools for Data Analysis and Data Science, First Edition. Daniel J. Denis.
© 2020 John Wiley & Sons, Inc. Published 2020 by John Wiley & Sons, Inc.

discussion of how $p < 0.05$ "happens." Assuming at least some difference in means or some correlation in the sample greater than zero, in other words, for at least some effect in the sample, we are guaranteed to reject the null hypothesis for sufficient sample size. That is, as sample size increases larger and larger, it is a mathematical fact that smaller and smaller p-values will occur. So how does all this play into power? Well, quite simply, if you have enough statistical power, and there is an effect in your sample, you will at some point reject the null hypothesis. **It's a guarantee**. This again is why as a researcher, you shouldn't place too much stock into p-values alone, since statistical significance in a given case might just mean a very large sample size was drawn yielding overwhelming power (that doesn't mean power or large samples are somehow "bad," a point we will return to later).

The level of statistical power is determined by four elements:

1) **Effect Size**. As mentioned, if there is at least some effect in your sample, however small, then at some point if you collect a large enough of a sample, you will reject the null hypothesis. But if the effect is very small, it may take an extremely large sample size before this rejection occurs. However, if the effect is very large to begin with, then a smaller sample will suffice to obtain statistical significance. Hence, one determinant of statistical power is how big the actual effect is that is being studied. All else equal, **larger effects yield greater statistical power**. In the case of a t-test, for instance, if the distance between means is large, then it will require fewer subjects than not to obtain statistical significance. Likewise, if the correlation in the population is large, it will require fewer subjects than not to demonstrate it. An analogy here will help. If I throw a rock into the water and seek to detect a "splash," that splash will be much more easily detected if the splash is large compared to if it is small. The "splash" (or amount of displacement of the water due to the rock) is the effect. Given that the effect is large, I will be able to detect it with ease. On the contrary, if the splash is minimal or hardly visible, that is, a quite small effect, I will have difficulty detecting it.

2) **Population Dispersion or Variability**. All else equal, if elements of the population are more similar than not (i.e. the case of low variability), then this provides a more powerful test than if elements are very different from one another. That is, for populations that are scattered and in which there is much **noise**, such will weaken the power of the test. That's why studying human subjects, especially in social studies, often requires relatively large sample sizes, since these populations are usually quite "noisy," and it's harder to detect an effect amongst the noise. On the other hand, in well-controlled animal studies, where there is typically assured to be very little variability among participants, it usually becomes much easier to detect effects since there is very little noise to wade

through. Experimenters usually purchase rats that are bred at the same time, born under identical conditions, and are assured they are as similar as possible before subjecting them to a control versus experimental group. This **similarity** among participants practically guarantees very little variability, so if the treatment does have an effect, it can usually be spotted with relatively few animals. However, we often cannot have such control with human subjects, and that vast variability among them unduly contributes to the noise factor in an experiment, requiring larger and larger samples to sometimes detect even minor effects of the independent variable.

3) **Sample Size**. As already mentioned, for a given effect, the larger the sample size, the greater the statistical power. Recall that this is also why a rejection of the null hypothesis is practically guaranteed for even small effects. If the effect is large, and population variability low, we often require very little in the way of sample size to demonstrate an effect. On the other hand, if the effect is small and hardly noticeable, we often require quite large sample sizes to reject the null hypothesis. If the effect is small and there is high variability among participants, we may require an overwhelmingly large sample size to detect that small effect.

5.2 Does That Mean Power and Huge Sample Sizes Are "Bad?"

Given our discussion of power, it may at first appear that having high degrees of power and very large sample sizes are not good things. After all, if a rejection of the null hypothesis is virtually guaranteed for very high power and huge sample sizes, it might at first glance appear that these are "negatives" and drawbacks to the research design, since they pretty much all the time lead to rejections of the null. However, this is not the case. **You can never have "too big" of a sample, and you can never have "too much" power**. What you can have is a misunderstanding of what $p < 0.05$ means and not knowing when to look at effect size, that's the real issue, not having a lot of power or sample size.

Remember, the only reason we compute statistics on small samples in the first place is because obtaining the actual population in most cases is very difficult, time-consuming, and expensive. If we could, we would always obtain population data and study population parameters directly. Hence, when you gather increasingly larger sample sizes, you are, in theory, getting closer and closer to obtaining the population data, which is the "ideal." Just because the p-value has shortcomings as a way of evaluating evidence and is largely a function of sample size, it does not mean we should amend the situation by not gathering large samples. What it does mean is that we should be very careful when interpreting *p*-values and recognize that we need to interpret them in conjunction with effect sizes for them to make any good **scientific** sense.

You cannot have "too much" power or "too big" of a sample size. You cannot have a test that is too powerful. Remember the goal in research is to learn something about population parameters; hence, in general, bigger samples are preferable to smaller ones. However, since we know that statistical significance (e.g. $p < 0.05$) can be a function of large samples, we simply need to be aware of this and turn to effect size to help us make meaning out of the scientific result. There does come a point where larger and larger samples will provide diminishing returns in terms of knowledge about the parameters, but that does not equate to bigger samples being unfavorable because of their influence on p-values.

5.3   Should I Be Estimating Power or Sample Size?

One question that often arises is whether one should estimate power for a given sample size, or sample size for a given level of power. In short, it doesn't really matter, since one is a function of the other. In any power calculation, if you specify sample size first, you can then estimate the corresponding statistical power. If you already know your desired level of power, then you can estimate how big of a sample size you will require to hit that mark. Usually, in planning a study, researchers will set power at a given level, then use that to determine how big of a sample size they will need. However, it is often the case in research contexts that the availability of a sample may be difficult to come by, and that the researchers know in advance the size they are limited by. So the researcher may ask, "Given I can only collect a sample size of 100, how much power will that afford me?" Then, if they learn that will only give them say, 0.60 power, they can brainstorm whether there is any feasible way of increasing sample size, or reducing variability. And if not, then they may side with not doing the study in the first place, or treat it as a **pilot study**. Pilot studies are those performed on typically very small sample sizes, with the understanding that formal inferences to the population will likely not be able to be made given the small sample size. However, if the pilot study is successful and yields a respectable effect size, it may encourage that same investigator or other investigators to follow up with conducting a more thorough study with larger sample sizes as to be able to make formal inferences.

So in sum, whether you estimate degree of power or size of sample won't really matter, as one is a function of the other. In our examples featured in this chapter, we will trade-off between the two approaches.

5.4 How Do I Know What the Effect Size Should Be?

5.4.1 Ways of Setting Effect Size in Power Analyses

Since the level of power will be determined in part by effect size, an obvious next question is how one should go about setting an effect size estimate. As we have discussed, all else equal, the search for larger effects typically requires less sample size, and hence when effects are fairly large, statistical power can be relatively high even for moderate sample sizes. So when conducting a power analysis, how should a researcher go about setting their effect size? There are a few ways, not necessarily always mutually exclusive, that a researcher can go about setting or anticipating effect size before conducting a study:

1) **What the researcher actually believes the effect size is in the population**. This is probably the most common way of setting effect size. The researcher simply sets effect size at what he or she believes is likely to be true in the population. Estimates obtained this way are often informed by the literature base or prior research, or, simply by knowledge of the researcher about the field in which he or she works. In many social studies, for instance, a researcher may be able to provide a good "guesstimate" that the true effect size will be in the neighborhood of 20–30% variance explained. In other studies that may have more experimental controls, estimates may range much higher. The point is that when effect sizes are estimated this way, they correspond to what the researcher believes is a **realistic estimate** of what will be found in the population. It is the researcher's "best guess" at what the effect will likely be.

2) A second way of fixing effect size in a power analysis is to set it at the **minimum effect that the researcher would deem important for the study**, such that any smaller number would make the study not worth doing. For example, in evaluating the effectiveness of a new medication on reducing headache symptomology, if the new medication would not result in an at least 10% reduction in headaches, for example, then doing the experiment may not be worthwhile to the investigator. Given this, the researcher may choose to fix effect size at 10% reduction between the control and experimental group. Now, that does not mean the true effect size will necessarily be found to be a 10% reduction. What it does mean is that finding anything less than 10% would not be of value to the researcher, hence 10% is set at the **minimal** effect worth discovering.

3) A third way, which is more or less a blend of the above two approaches, is to **set effect size very conservatively**. That is, set it at a very small number to practically guarantee adequate power and sample size. This way, the researcher guards against the possibility of the "worst case scenario" in which effect size is even smaller than he or she thought it would be, yet should that event occur, adequate sample size and power are still assured as a result of the ultraconservative estimate. This is by far the most cautious approach as it concerns setting effect size, as it practically guarantees that for a given level of power, an adequate sample size will be recruited. Why? Because the presumed effect is so low, we are planning for the worst case already when estimating sample size. For example, even if a researcher might believe the true effect is in the neighborhood of 10% variance explained, he or she may set the effect size at 5% just to guard against the possibility of discovering an effect much smaller. If she set it at 10% for the power analysis, and the research revealed an effect of less than 10%, she may not have the power to reject the null. Hence, a useful strategy is to set small effect sizes, and then hope they come out to be larger than that once we conduct the study. However, if they do not, then with this conservative approach we still have a shot at rejecting the null with sufficient power.

In practice, effect sizes are usually set by a sort of **mixture** of the above approaches. However, if in doubt, it is generally recommended that setting a fairly small and conservative estimate is best, as to ensure recruiting a minimally sufficient sample size.

5.5 Power for *t*-Tests

Our first example of power is that for the simple between-subjects *t*-test. We can use the function `pwr.t.test()` in the package **pwr** to conduct a variety of power analyses for different *t*-test situations. The fields we will need to enter are the following:

```
pwr.t.test(n = , d = , sig.level = , power = , type =
c("two.sample", "one.sample", "paired"))
```

- n is sample size
- d is effect size (Cohen's *d*)
- sig.level is the level of significance desired for the given test (e.g. 0.05, 0.01)
- power is the level of statistical power desired
- type is the kind of *t*-test that we are running, options include the two-sample independent groups *t*-test, the one-sample test, and the paired-samples test.

5.5.1 Example: Treatment versus Control Experiment

Suppose a scientist plans a study in which she would like to investigate the influence of a treatment for staying awake longer. She hopes that people who take the stimulant will be more alert and will need less sleep than those who do not take it. The control group will receive no treatment, while the experimental group will receive the stimulant. Suppose the researcher sets the level of significance at 0.05, desires a level of power of 0.90, and anticipates an effect size of Cohen's *d* of 0.5. Given these parameters, the scientist is interested in estimating required sample size. Estimated sample size can be computed as follows:

```
> library(pwr)
> pwr.t.test(n = , d = 0.5, sig.level = 0.05, power = 0.90,
type = "two.sample")

     Two-sample t test power calculation

             n = 85.03128
             d = 0.5
     sig.level = 0.05
         power = 0.9
   alternative = two.sided

NOTE: n is number in *each* group
```

The output given by R is more or less self-explanatory. R lists the parameters we specified (i.e. d = 0.5, sig.level = 0.05, power = 0.90), and notes that it is a two-sided test (typically, in most research situations you always want to estimate power for two-sided alternatives). The number of required participants for this project is 85.03 per group. That is, we require approximately 85 participants per group to run this experiment for a level of power of 0.90 and anticipated effect size of $d = 0.5$.

For demonstration purposes, let's consider what would happen if we increased our effect size to say, 2.0. That is, suppose the researcher, instead of anticipating an effect of $d = 0.5$, anticipated one of 2.0. If you are understanding power correctly, then you can pretty much predict what will happen to the ensuing sample size estimate. Recall that as effect size gets larger, all else equal, we require lesser of a sample size to detect it. Having now anticipated our new effect size to be $d = 2.0$, but keeping the significance level and power level the same, we now re-estimate sample size:

```
> pwr.t.test(n = , d = 2.0, sig.level = 0.05, power = 0.90,
type = "two.sample")
```

```
Two-sample t test power calculation

              n = 6.386753
              d = 2
      sig.level = 0.05
          power = 0.9
    alternative = two.sided
```

```
NOTE: n is number in *each* group
```

Wow! Notice that for a whopping effect size of $d = 2.0$, we require only approximately six subjects per group! When you're expecting large effects, you simply do not require much of a sample size to detect it. Since the anticipated or desired effect of 2.0 is much greater than 0.5, it stands we do not require as much of a sample size to "spot" it. Hence, if the scientist really believed the effect would be this large, then she could theoretically get away with only collecting about six subjects per group to demonstrate the effect. In passing, it is worth mentioning that whenever you obtain a sample size estimate with decimal places, it is always good practice to round up as to provide the most conservative estimate. For our example above, the number of participants called for is 6.386753 per group. Of course, collecting this many participants is impossible as it is impossible to collect a fraction of a participant. Should we round up or round down? **Round up if you desire the most conservative estimate**. For our example, we would round up to 7. The reason for this, again, is that rounding up provides a bit more of a conservative estimate than rounding down. That is, proceeding as if you require a bit more of a sample size than you may actually need guards against having insufficient sample size.

5.5.2 Extremely Small Effect Size

Now, again for demonstration, using the same example as above, let's assume that instead of $d = 0.5$ or $d = 2.0$, the scientist anticipated an extremely small effect size. Let's suppose $d = 0.1$, which is considered to be quite small in most research contexts. We will leave all other parameters the same. For such a small effect, leaving all other parameters at the levels formally set, our new computation of power would be:

```
> pwr.t.test(n = , d = 0.1, sig.level = 0.05, power = 0.90,
type = "two.sample")
```

```
    Two-sample t test power calculation

              n = 2102.445
              d = 0.1
      sig.level = 0.05
```

```
        power = 0.9
  alternative = two.sided
```

NOTE: n is number in *each* group

Wow again! To detect such a small effect at $d = 0.1$, we require a whopping **2102 (or rounded to 2103) subjects per group** for the same level of power (i.e. 0.90). While this may at first glance seem surprising, it is entirely reasonable. Since there is such a small anticipated difference, we require many more subjects to spot the effect. That is, the effect of $d = 0.1$ is extremely small, it's barely noticeable, to use our analogy, **the rock is barely making a splash**. To detect that splash then, we will require a highly sensitive microscope in the form of a larger sample size to spot such a small effect.

However, as discussed earlier, we usually can't just invent at random what we would like or wish to see for effect sizes in populations. That is, there is a scientific reality as to what the effect size will be or will not be, and hence in estimating power, it is vital that the researcher **intelligently estimates sample size correctly**, either by what is a reasonable estimate of what he or she anticipates seeing, or the minimum effect desired in order to consider the experiment or study worthwhile. In our examples here, where we are actively manipulating the effect in the form of changing Cohen's d from 2.0 to 0.1 for instance, this is simply for demonstration. If power analyses were that simple such that researchers could enter effects of their dreams, then a researcher would always just enter a large effect size and rejoice in not having to collect large sample sizes! This, of course, would defeat the purpose of the scientific investigation, because once the experiment is performed, and the effect turns out to be much smaller, the researcher will be unable to reject the null hypothesis. That is, it pays no dividends for the researcher to pretend or hope for large or unrealistic effect sizes in a power analysis if they do not adequately represent a realistic estimate of the effect in the population.

Though at first glance it may seem desirable to input large effect sizes for a given level of power as to require fewer subjects, anticipating unrealistic effect sizes defeats the purpose and will nonetheless likely generate an experiment with low power. Effect size must be anticipated intelligently and realistically, either by inputting the effect that is likely to occur in the population, or inputting the minimal effect desired to make the experiment or study worthwhile. For instance, inputting a Cohen's d = 2.0 in the case of a t-test when in reality the more realistic anticipated effect is d = 0.5, is an exercise in fooling oneself, and using d = 2.0 in this case will simply generate an experiment with low power once it is found that the true effect is much smaller. Setting effect sizes conservatively will guard against this possibility.

5.6   Estimating Power for a Given Sample Size

Having above estimated sample size for a given level of power, we now instead estimate power for a specified sample size. In this instance, the same scientist may know in advance that she'll have a maximum of 100 subjects available, and would like to know how much power that would afford her in conducting her statistical test. Many times research investigators may be limited by how much of a sample they can draw, and hence it may be known in advance that 100 participants is all she will be able to realistically recruit. Or, she may be interested in simply obtaining a rough estimate of whether collecting 100 participants will yield adequate power.

For this computation then, we enter a value for sample size, but leave power = , open to be estimated:

```
> pwr.t.test(n = 100, d = 0.5, sig.level = 0.05, power = ,
type = "two.sample")

        Two-sample t test power calculation

              n = 100
              d = 0.5
      sig.level = 0.05
          power = 0.9404272
    alternative = two.sided
```

NOTE: n is number in *each* group

We can see that for a sample size of 100 per group, with all other parameters staying the same as in our previous example (i.e. $d = 0.5$, sig.level = 0.05, and for two-sided alternative), power is estimated at approximately 0.94. Be sure to note again that whether one chooses to estimate power or sample size, the end result will be the same. In most cases, however, because sample size is a relatively flexible variable, researchers will often prefer to specify desired power for a given effect size, then estimate how much of a sample they will require to detect such an effect. However, as mentioned, if sample recruitment is limited, or sometimes samples are gathered from **historical archival data** so that the researcher has absolutely no control over its size, inputting sample size first then estimating power is sometimes done.

5.7   Power for Other Designs – The Principles Are the Same

Above we have only featured the case of a *t*-test, but the point of our rather lengthy discussion of these examples has been to allow us to survey the issues and determinants of statistical power with real numbers and real estimates.

In your own research, the kind of test you may perform may change (e.g. you may perform a one-sample test, for instance, or any other kind of statistical test), but the issues we have surveyed and the relationships among determinants to power will remain the same. The goal shouldn't be then to memorize how to compute power for every design or experimental set-up out there. The goal should be rather to understand what power is, and how it is determined by such things as sample size and effect size, so that when you need to perform a power analysis in your own research, even if you have to look up the exact specifics on how to do it for the given research design, you are already familiar with the principles at play. **Always seek to learn the most general principles of whatever you are learning so that you can generalize what you have learned to new settings, analogous to how a musician learns chord–scale relationships and the "big picture" of how music works, then "new music" is usually a special case of those wider principles**. Good jazz musicians do not memorize a billion different elements of jazz. Instead, they learn the principles that underlie jazz music, and then adapt and recognize these commonalities in the songs and progressions they come across.

 Always seek to learn the most general principles behind whatever you are learning, as the commonalities behind power analyses and statistical techniques in general are much more prevalent than the peculiarities of a given approach. You can memorize 100 different things and still not understand any of them, or you can learn 1 thing, understand it well, and then apply that understanding to 100 different things.

Having demonstrated power principles through a *t*-test, we now briefly survey these same principles in a one-way ANOVA.

5.7.1 Power for One-Way ANOVA

We will study the analysis of variance (ANOVA) model in the following chapter, so if you are not familiar at all with ANOVA, it may be worth jumping ahead and reading that chapter before continuing on. However, if you already do have some experience with ANOVA, then the following discussion will already make good sense.

Estimating power for a one-way ANOVA is about as easy as for a *t*-test. However, the inputs will be a little bit different. To estimate power for a one-way between-subjects ANOVA, here's what we require:

```
pwr.anova.test(k = , n = , f = , sig.level = , power = )
```

In the above, k is the number of levels on the independent variable. Note that this does **not refer to the number of independent variables**, as that number is assumed to be one in this case because this is power estimation for a one-way ANOVA. The parameter k is the number of levels on that single independent variable. For example, if your independent variable is that of medication, and you are interested in testing different levels of it, such as 0.1, 0.2, 0.3 mg, then k refers to these different levels, which in this case would equal 3.

The value of f here requires some explanation. These values were defined by Cohen (1988) in his book on power analysis, and are meant as a general measure of **effect size**, analogous to how *d* was specified as a measure of effect in the *t*-test. The value of f is best understood in relation to a much more popular statistic, which is the well-known R^2 statistic. Since researchers habitually report and use R^2 statistics, thankfully there exists an easy way to convert from f to R^2:

$$f^2 = \frac{R^2}{1 - R^2}$$

Hence, to input a value for f in our power estimation, it's much easier to simply estimate a value for R^2, then do the conversion into f^2, then take the square root of f^2 to get f. For example, for a given R^2 of 0.20, the corresponding f^2 is equal to:

$$f^2 = \frac{0.20}{1 - 0.20} = \frac{0.20}{0.80} = 0.25$$

Hence, the value of *f* is therefore equal to $\sqrt{f^2} = 0.50$. This number of 0.50 then is the value that we would enter for f = , if we anticipated an R^2 value of 0.20. Now we can use this estimate of effect size in our ensuing power computation. For example, let's compute required sample size for power equal to 0.90 and level of significance of 0.05 for an ANOVA having k = 5 levels, again setting f = 0.5, which as we just noted corresponds to an R^2 value of 0.20:

```
> pwr.anova.test(k = 5, n = , f = 0.5, sig.level = 0.05,
power = 0.90)
        Balanced one-way analysis of variance power
calculation
                k = 5
                n = 13.31145
                f = 0.5
        sig.level = 0.05
            power = 0.9

NOTE: n is number in each group
```

Table 5.1 $R^2 \to f^2 \to f$ conversions.

R^2	f^2	f
0.10	0.11	0.33
0.20	0.25	0.50
0.30	0.43	0.65
0.40	0.67	0.82
0.50	1.00	1.00
0.60	1.50	1.22
0.70	2.33	1.53
0.80	4.00	2.00
0.90	9.00	3.00
0.99	99.00	9.95

For the specified parameters, we require approximately 13 subjects (or rounded to 14) in each of the 5 groups (i.e. representing the 5 levels of the independent variable) to achieve power of approximately 0.90 for an effect size of $R^2 = 0.20$.

5.7.2 Converting R^2 to f

Instead of having to perform the conversion from R^2 to f each time we would like to perform a power analysis, it would be much more convenient to simply compute them beforehand and be able to refer to the conversion as needed. This is exactly what Table 5.1 depicts, which is adapted from Denis (2016). Table 5.1 contains conversions from R^2 to f for a variety of effect size magnitudes.

Immediately noticeable from Table 5.1 is that as the value of R^2 increases from 0.10 to 0.99, the values for f^2 and f increase accordingly. However, the increase is not perfectly linear.

5.8 Power for Correlations

Next, we consider the case of estimating power for correlations, specifically the Pearson product-moment correlation coefficient. To estimate power for a Pearson correlation coefficient, we require the following inputs:

```
pwr.r.test(n = , r = , sig.level = , power = )
```

In the above, n is again the sample size, r is the correlation coefficient anticipated in the population, that is, it is the correlation we are seeking to detect with our test. As an example, suppose a researcher wishes to detect a correlation of $r = 0.10$ at a significance level of 0.05, and sets power equal to 0.90. The estimated number of subjects required for these inputs is computed by:

```
> pwr.r.test(n = , r = .10, sig.level = 0.05, power =0.90)
```

```
     approximate correlation power calculation (arctangh
transformation)

           n = 1045.82
           r = 0.1
   sig.level = 0.05
       power = 0.9
 alternative = two.sided
```

We see that to detect a correlation of size $r = 0.10$ in the population at a level of power equal to 0.90, upward of 1046 (rounded up from 1045.82) participants is required! This may seem like an overwhelming number of participants, but sample size requirements for correlation coefficients can be exceedingly high, higher than for even relatively complex regressions.

Now, suppose instead that the correlation we are seeking to detect was much larger. Suppose $r = 0.90$, for the same level of power (0.90). Would we expect sample size requirements to increase or decrease? If you're understanding power, you'll recall that the larger the effect size, in this case, r, the less number of subjects needed to detect that effect for a given level of power. Thus, if our intuition is correct, we should require much fewer participants for these input parameters. Let's confirm that our intuition is correct:

```
> pwr.r.test(n = , r = .90, sig.level = 0.05, power =0.90)
```

```
     approximate correlation power calculation (arctangh
transformation)

           n = 7.440649
           r = 0.9
   sig.level = 0.05
       power = 0.9
 alternative = two.sided
```

Our intuition was spot on. To detect an $r = 0.90$, we require only approximately 7–8 subjects. Once more, this demonstrates the influence of size of presumed effect on sample size requirements for a given level of power. Again, if the effect is presumed to be very large, and thus "easily detectable," then all else equal, we require a much smaller sample size than if the effect is thought to be very small and thus difficult to spot.

5.9 Concluding Thoughts on Power

In this chapter, we surveyed power analysis through *t*-tests, ANOVA, and correlations, demonstrating the essentials of how estimating sample size and power works. These are just a few examples of power computations using R; however, they merited a chapter in their own right to provide a fairly thorough discussion and demonstration of the determinants of statistical power. The principles studied in this chapter generalize to a wide variety of designs and statistical models, and hence if you master the relationships between power, effect size, sample size, etc. then you will be well on your way to understanding power analysis overall and how it relates to *p*-values and the establishment of statistical evidence in the social and natural sciences. And if you have so-called "Big Data," then while in most cases you are still technically inferring population parameters (i.e. no matter how large our data sets, we usually never truly have complete populations), power isn't really a concern anymore because you implicitly have sufficient power already. **Both statistically and scientifically, hypothesis testing still takes place on those parameters.** The only difference is that *p*-values are generally meaningless since the null will always be rejected. But this is the same principle for even much smaller sample sizes, though still large enough to make *p*-values arbitrary from a scientific point of view. Don't let anyone tell you "Big Data" is making hypothesis-testing "obsolete" somehow, because that is clearly not the case, both from statistical and scientific perspectives.

Key points to take away from this chapter include:

- Power is the probability of detecting a false null hypothesis. One cannot have "too much" power. However at some point, increasing sample size to achieve greater power is not worth the investment of collecting such a large sample. Once a respectable level of power has been achieved, collecting an increasingly larger sample size usually affords no advantage. However, it stands to reason that one cannot have "too big" of a sample either. Always remember that the overriding purpose of research is to study populations, not samples, and hence larger samples are usually preferred over smaller ones. But as mentioned, the cost-effectiveness of collecting increasingly larger samples may not be worth it, however, if the same conclusions can be drawn from much smaller samples.
- The larger the presumed effect, all else equal, the smaller the sample size you require to detect that effect. Large effects require less sample size than smaller effects.

Exercises

1   Define and discuss the nature of statistical power. What is power, and why is it important or relevant to research and statistical estimation?

2   Explain how for even a very small effect size in the sample, the null hypothesis can always be rejected for sufficient sample size. Give a research example to demonstrate your understanding.

3   Discuss the determinants of statistical power, by explaining the effect that manipulating each one has on the degree of power achieved.

4   Why is power typically computed before an experiment is executed?

5   Discuss the ways of estimating effect size in the population. Which way is best in ensuring sufficient power for an experiment or study?

6   For an R^2 of 0.20, what is the corresponding value for f?

7   Demonstrate in R the effect of an increasing effect size estimate when estimating power for detecting a correlation. Set power at 0.80, sample size at 500, and use a traditional significance level of 0.05.

8   Your colleague says to you, "Be careful not to have too much power, it will invalidate your results." Correct your colleague.

9   Your colleague performs an experiment with a sample size of 10,000 participants, yet fails to reject the null hypothesis. Comment on what you believe the effect size likely is for the experiment.

10   How does the concept of statistical power fit in with the bigger picture and advent of "Big Data"? Does "Big Data" invalidate the relevance or need for hypothesis-testing and parameter estimation? Why or why not?

6

Analysis of Variance

Fixed Effects, Random Effects, Mixed Models, and Repeated Measures

---

LEARNING OBJECTIVES

- Understand the logic of how analysis of variance (ANOVA) works.
- How to verify assumptions in ANOVA with both inferential tests and exploratory graphs.
- How to distinguish between fixed effects, random effects, and mixed models ANOVA.
- How to use post hoc tests such as Tukey's HSD to evaluate pairwise comparisons after a statistically significant F-stat.
- How to use both `aov()` and `lm()` to run ANOVA in R.
- How to understand the concept of an interaction in ANOVA and read interaction graphs.
- How to perform simple effects analyses in R to follow up in the presence of an interaction.
- How to use and interpret statistics such as AIC and BIC in ANOVA models.
- How to run random effects and mixed models in R.
- How to distinguish repeated-measures models from between-subjects and run these models in R.

The analysis of variance (ANOVA), next only to regression analysis, is undoubtedly the most common of statistical methods used in applied research. There are two ways to understand ANOVA: the first is as an extension of the univariate t-tests of earlier chapters, the second is as a special case of the wider regression analysis model to be surveyed in a later chapter. Since we have not yet studied regression, we will first introduce ANOVA as an extension of the t-test, then revisit our second formulation once we have studied regression.

Univariate, Bivariate, and Multivariate Statistics Using R: Quantitative Tools for Data Analysis and Data Science, First Edition. Daniel J. Denis.
© 2020 John Wiley & Sons, Inc. Published 2020 by John Wiley & Sons, Inc.

6.1 Revisiting *t*-Tests

Recall in the independent-groups *t*-test, we tested null hypotheses of the sort

$$H_0 : \mu_1 = \mu_2$$

against an alternative hypothesis of the kind,

$$H_1 : \mu_1 \neq \mu_2.$$

To understand ANOVA, we need to first survey and unpack the equations for the *t*-test. The *t*-test we performed to evaluate the null hypothesis compared sample means in the numerator to a standard error of the difference in means in the denominator:

$$t = \frac{\bar{y}_1 - \bar{y}_2}{\sqrt{\dfrac{s_1^2}{n_1} + \dfrac{s_2^2}{n_2}}}$$

where *t* was evaluated on $(n_1 - 1) + (n_2 - 1)$ degrees of freedom. When sample sizes are unequal, $n_1 \neq n_2$, we pool variances and perform the following *t*-test instead:

$$t = \frac{\bar{y}_1 - \bar{y}_2}{\sqrt{s_p^2 \left(\dfrac{1}{n_1} + \dfrac{1}{n_2} \right)}},$$

where s_p^2 represents a pooled variance, equal to $s_p^2 = \dfrac{(n_1 - 1)s_1^2 + (n_2 - 1)s_2^2}{n_1 + n_2 - 2}$.

The above *t*-tests work fine so long as we have only two means in our experiment or study. However, when we have three or more means, a simple *t*-test is insufficient for evaluating mean differences. For example, instead of simple hypotheses such as $H_0 : \mu_1 = \mu_2$, suppose we wished to evaluate hypotheses of the sort,

$$H_0 : \mu_1 = \mu_2 = \mu_3 = \mu_4,$$

where we now have 4 means. There is no way a univariate *t*-test could evaluate this null, unless of course we chose to perform a bunch of *t*-tests for each mean comparison possible. For example, we could, in theory, compare μ_1 to μ_2, then μ_1 to μ_3, then μ_1 to μ_4. Then, we could compare μ_2 to μ_3, and so on for the remainder of the means. What's wrong with this strategy? Recall that with each comparison is a risk of making a type I error, typically set at 0.05. Thus, if we pursued this approach, we would be unduly inflating the overall type I error rate across the **family of comparisons** we wished to make. Instead of minimizing our risk of error, we would be inflating it across numerous *t*-tests. This is unacceptable, and we desire a new approach to evaluate the null hypothesis of several means. Not only will our new approach address the type I error issue, but it will also lend itself to being a much more complete conceptual and statistical model than could ever be possible with only univariate *t*-tests.

6.2 Introducing the Analysis of Variance (ANOVA)

6.2.1 Achievement as a Function of Teacher

The best way to understand the principles behind ANOVA is to jump in with a simple example, and at each step along the way, seek to understand the logic of the procedure right up to analyzing data in R. In this way, we work with an actual data set, and can see the principles behind ANOVA at work as we proceed.

In our example, suppose we wished to evaluate a null hypothesis that achievement population means are the same across a set of pre-selected teachers. That is, as a researcher, you wish to know whether achievement of students differs depending on which teacher they are assigned. In this example, you use teachers of your particular choosing as levels of the independent variable. It is vital that **you selected the specific different teachers** for this example rather than they be **randomly** sampled. That is, any results you obtain you wish to generalize only to the teachers you selected, and no others. This, as we will elaborate on later, is what makes teacher a **fixed effect**. Our null hypothesis is thus:

$$H_0 : \mu_1 = \mu_2 = \mu_3 = \mu_4$$

The alternative hypothesis is that somewhere among the means, there is at least one mean difference of the sort,

$$H_1 : \mu_1 \neq \mu_2, \mu_3 \neq \mu_4, \text{ etc.}$$

A few things to remark about our alternative hypothesis. First, it is very **nonspecific**. It is simply a hypothesis that somewhere among the means, there is a difference, but we are not hypothesizing exactly where that difference may be. It could be between the first two means, or maybe the second two means, etc. This is what we mean by the alternative being very nonspecific. It stands that if we reject the null hypothesis, then we will know that at least somewhere among the means, there is a mean difference, but we will need to perform further tests in order to learn where those differences may be. We will return to this point later.

The hypothetical achievement data for this example appears in Table 6.1.

It is always important to inspect the data somewhat even at a visual level before subjecting it to even exploratory analyses, not to mention inferential. We look at the data in Table 6.1 and make the following observations:

- We note that **within** each teacher grouping, there definitely does appear to be variability. That is, scores within groups differ from one another. For example, within teacher 1, we see achievement scores ranging from 65 as a minimum to

Table 6.1 Achievement as a function of teacher.

Teacher			
1	2	3	4
70	69	85	95
67	68	86	94
65	70	85	89
75	76	76	94
76	77	75	93
73	75	73	91
$M = 71.00$	$M = 72.5$	$M = 80.0$	$M = 92.67$

76 as a maximum. Likewise, within teacher 2, we see scores ranging from 68 as a minimum to 77 as a maximum.

- We note that **between** teacher groupings is also exhibited a fair degree of variability. For instance, we notice that as we move from teacher 1 to teacher 4, scores appear to generally increase. A useful summary statistic to measure or "capture" this average increase across groups is the **sample mean**, computed for each group, yielding 71.0, 72.5, 80.0 and 92.7 for teachers 1 through 4, respectively.
- The key point so far is that we have noticed in Table 6.1 that we have both **between** and **within** variability. These are two **sources of variability** that we will wish to compare in some fashion as we move forward.

Now, we fully expect that sample means should differ by teacher to some degree even if the null hypothesis were true. Pause and think about that for a moment. Do you agree? That is, even if the null hypothesis of $\mu_1 = \mu_2 = \mu_3 = \mu_4$ were "true," which is about **population means**, do you really think **sample means** would also be exactly equal? Likely not. Why not? Because of intrinsic variability. Even under a null hypothesis of equal population means, we rarely if ever expect sample means to also be equal. So when we look at the sample means in Table 6.1, we fully expect them to be different to some degree even if the null hypothesis is not rejected. The question we wish to ask, however, is whether the differences in sample means is large enough relative to within variation under the null hypothesis such that we could conclude actual population mean differences. That is an analogous question that we asked when dealing with a coin in the binomial, or in t-tests and other tests earlier in the book. Is the difference we are seeing in our sample large enough relative to variation we would expect to see under the null hypothesis that we are willing to bet there is a difference in means in the population? That is the question we are asking, not whether there are sample mean differences. Again, we fully expect sample means to be different. The question is whether between-group variation is large enough relative to within-group variation that we can confidently reject the null hypothesis $H_0 : \mu_1 = \mu_2 = \mu_3 = \mu_4$. The question of interest is about **population means**, not sample means.

Entered into R, our data appear below:

```
> achiev <- read.table("achiev.txt", header = T)
> achiev
   ac teach text
1  70     1    1
2  67     1    1
3  65     1    1
4  75     1    2
5  76     1    2
6  73     1    2
7  69     2    1
8  68     2    1
9  70     2    1
10 76     2    2
11 77     2    2
12 75     2    2
13 85     3    1
14 86     3    1
15 85     3    1
16 76     3    2
17 75     3    2
18 73     3    2
19 95     4    1
20 94     4    1
21 89     4    1
22 94     4    2
23 93     4    2
24 91     4    2
```

Notice in our data frame is also a "text" variable. We ignore this variable for now, and will use it a bit later in the chapter. For now, we are only interested in variable ac for "achievement" and teach for "teacher."

We first obtain a few boxplots of achievement by each level of the teach factor:

```
> attach(achiev)
> boxplot(ac ~ teach, data = achiev, main="Achievement as
a Function of Teacher",
+ xlab = "Teacher", ylab = "Achievement")
```

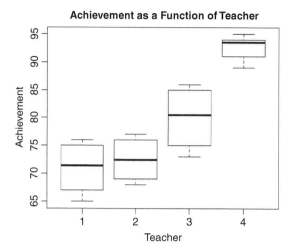

We note from the boxplots that **median** achievement appears to increase as we move from teacher 1 to teacher 4. The medians for teachers 1 and 2 appear to be quite similar, while the median for teacher 4 is the highest of all teachers. These are not means, though they give a fair indication of the between-group spread referred to earlier.

6.3 Evaluating Assumptions

Recall that virtually every statistical model comes with it a set of assumptions that if not satisfied, can severely compromise what we can conclude regarding obtained *p*-values computed on parameter estimates. For example, if assumptions are rather severely violated, then generated *p*-values from the ANOVA may not be accurate, even though software still reports them. A *p*-value equal to 0.045 for instance, is only as "good" as the assumptions of the statistical model on which it is based. For this reason, verifying assumptions is usually necessary for most statistical models. Many violations of assumptions can be tolerated to some extent, but a check on as many assumptions as possible is always good form for good data analysis. The consequences of violating certain assumptions is more catastrophic than violating others, as we will soon discuss.

A first assumption that ANOVA holds is that within each population, distributions are more or less **normal** in shape. The assumption of **normality of population distributions** can be evaluated by plotting histograms or other similar visual displays for data within each population when adequate sample size per group is available. For our data, since we have only six observations per cell, attempting to verify this assumption of normality per group is virtually impossible. Nonetheless, let's take a look at the distributions of achievement for each level of the teach factor for the sake of demonstration. For this we will use the package **FSA**, while also generating a factor out of teach. This is accomplished by factor(teach):

```
> library(FSA)
> f.teach <- factor(teach)
> hist(ac~f.teach, data = achiev)
```

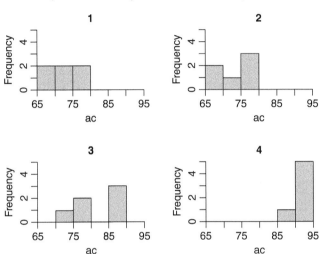

As we can see, with only six observations in each teach group, normality doesn't really have a chance at holding true. However, with more dense distributions, plotting histograms such as this would be useful to inspect distributions at each level of the independent variable.

6.3.1 Inferential Tests for Normality

In addition to graphical displays, we can also conduct formal inferential tests to evaluate normality, should we choose. One such test that can be performed is the **Shapiro–Wilk normality test** (we surveyed this test earlier in the book), which we will demonstrate on all of the achievement data in entirety to simply get a sense of overall normality across all groups (if we had sufficient sample size per group, we would perform the test individually for each group):

```
> shapiro.test(ac)

        Shapiro-Wilk normality test

data:   ac
W = 0.90565, p-value = 0.02842
```

The value of the statistic W is equal to 0.90565, and its associated p-value is equal to 0.02842, which is statistically significant. Since the null hypothesis is that the population distribution is normal in shape, we reject the null hypothesis and **conclude it is not normal**. Now, should this result cause us alarm? Not really. As mentioned, the assumption of normality refers in general to normality within each group, not normality of the data as a whole. However, with only six observations per group, realistically evaluating normality per group is hopeless. The test we just ran simply gives us a sense of whether overall, the distribution of scores appears to at least approximate in some sense a normal distribution. A plot of the corresponding histogram reveals data that are not normally distributed, but nonetheless shows sufficient variation of `ac` scores:

```
> hist(ac)
```

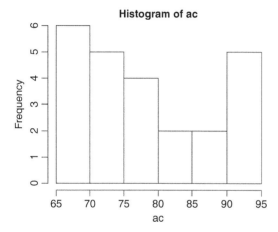

6.3.2 Evaluating Homogeneity of Variances

The analysis of variance procedure also assumes that population variances in each level of the independent variable are equal. Of course, again, we fully expect sample variances to differ to some extent, but the question as always is whether sample differences are large enough to give us reason to suspect population variances are unequal. The null hypothesis we wish to evaluate is:

$$H_0 : \sigma_1^2 = \sigma_2^2 = \sigma_3^2 = \sigma_4^2,$$

where σ_1^2 is the variance for level 1, σ_2^2 the variance for level 2, etc. In evaluating the assumption of homogeneity of variances, R provides several options. These include the **Fligner–Killeen test**, **Bartlett's test**, and **Levene's test**. The differences between these tests are beyond the scope of this book. For our data, we choose the Fligner–Killeen test. This is a nonparametric test that is robust to violations of normality and comes highly recommended by some (e.g. see Crawley, 2013):

```
> fligner.test(ac~f.teach, data = achiev)

Fligner-Killeen test of homogeneity of variances
data:   ac by f.teach
Fligner-Killeen:med chi-squared = 10.813, df = 3, p-value =
0.01278
```

The obtained *p*-value is equal to 0.01278, providing evidence to reject the null hypothesis at a preset level of significance of 0.05. However, if we used a more stringent level such as 0.01, the result is no longer statistically significant. So what do to? We are probably safe to continue with the ANOVA without any further adjustments, as ANOVA is quite robust to violations of unequal variances, especially for cell sizes that have equal numbers of observations. Though adopting "rules of thumb" in statistics is not generally advised in most circumstances since there are always exceptions to them, a rule of thumb happens to work well with regard to variances, and is one of the only rules of thumb given in this book. As an informal guideline then, so long as one variance is no larger than 4–5 times another variance, assuming equal sample size per group or cell, it is usually safe to proceed with the ANOVA without any further adjustment. ANOVA is less robust for unequal cell sizes. Let's take a look at the variances for our data by each level of the teach factor:

```
> aggregate(ac ~ f.teach, FUN = var)
  f.teach           ac
1        1 19.600000
2        2 15.500000
3        3 35.200000
4        4  5.066667
```

The largest variance is in level 3 of teach, equal to 35.2, while the smallest is in level 4 of teach, equal to 5.07. Hence, since the largest variance is more than a factor of 4 or 5 times the smallest, we would appear to have a violation of variances and may prefer to interpret a more robust test in the ANOVA called the **Welch test**. We will explore that test briefly when we conduct the ANOVA, and compare the results to an unadjusted ANOVA that assumes equal variances. As well, the `aggregate()` function used above can be used to obtain other statistics as well.

For instance, suppose we wanted to obtain the sample means (instead of variances) by group:

```
> aggregate(ac ~ f.teach, FUN = mean)
f.teach        ac
1         1 71.00000
2         2 72.50000
3         3 80.00000
4         4 92.66667
```

6.4 Performing the ANOVA Using `aov()`

We now demonstrate the ANOVA using the `aov()` function in R. For our ANOVA, because `teach` is a factor, we need to treat it as such in R. That is, we will ask R to create a factor variable out of `teach`:

```
> f.teach <- factor(teach)
> f.teach
[1] 1 1 1 1 1 1 1 2 2 2 2 2 2 3 3 3 3 3 3 4 4 4 4 4 4
Levels: 1 2 3 4
```

Though we have already obtained the means of achievement by teacher, we do so once more, this time using the `tapply()` function, yet another way to get summary statistics by group:

```
> tapply(ac, f.teach, mean)
        1        2        3        4
71.00000 72.50000 80.00000 92.66667
```

We confirm that the above means are the same as computed in Table 6.1. Using `tapply()`, we could have also obtained other statistics of interest, such as the median:

```
> tapply(ac, f.teach, median)
   1    2    3    4
71.5 72.5 80.5 93.5
```

The so-called "grand mean" of the data is, in this case, the mean of the group means (i.e. $[71.0 + 72.5 + 80 + 92.67]/4$), which we can compute in R as

```
> mean(ac)
[1] 79.04
```

The caveat "in this case" above refers to the fact that we have a **balanced design**, meaning that we have equal numbers of participants in each group.

Had the data been **unbalanced** (for instance 5 individuals in one group, and 6 in the others), then the grand mean (sometimes called the **overall mean** in the case of unequal sample sizes per group) of the data would have been equal to the **mean of all data points** rather than the mean of unweighted means.

6.4.1 The Analysis of Variance Summary Table

In every ANOVA there is a so-called "Summary Table" that is generated, which depicts how the variation for the given data has been partitioned into between versus within. This partition takes place as a sum of squares for each source of variation. Hence, in most software output, you will see something analogous to SS between and SS within that reveals the partition. These sums of squares are then divided by degrees of freedom to obtain what are called the **mean squares**.

To run the ANOVA in R:

```
> anova.fit <- aov(ac ~ f.teach, data = achiev)
> summary(anova.fit)
            Df Sum Sq Mean Sq F value   Pr(>F)
f.teach      3 1764.1   588.0   31.21 9.68e-08 ***
Residuals   20  376.8    18.8
---
Signif. codes: 0 '***' 0.001 '**' 0.01 '*' 0.05 '.' 0.1 ' ' 1
```

We summarize the output from the summary table:

- f.teach is the source of variation attributable to teacher differences; it has 3 degrees of freedom associated with it (i.e. the number of groups for teacher minus 1), and a sum of squares value of 1764.1. Its mean square is computed by dividing SS by df, that is, $1764.1/3 = 588.0$. ANOVA will use the mean squares to compute the corresponding F ratio (which we will describe shortly). Eta-squared, η^2, which is a measure of **effect size** for the ANOVA, is equal to SS teach/(SS teach + SS residuals) = $1764.1/(1764.1 + 376.8) = 0.82$. This is interpreted as approximately 82% of the variance in achievement is accounted for by teacher differences.
- Residuals represents the source of variation left over after consideration of teacher differences. That is, it's the "error" source of variation that remains unexplained after modeling teacher differences. It has associated with it 20 degrees of freedom, since there are 6 observations per group, and 1 degree of freedom per group is lost, leaving us with 20 for the overall ANOVA. To get the corresponding mean square, we divide 376.8 by 20, yielding a mean squares value of 18.8.

- As mentioned above, mean squares (Mean Sq) were computed next for both f.teach and Residuals. To get the corresponding F value, we divide MS for f.teach by MS for Residuals, that is, 5.88.0/18.8 = 31.21. Under the null hypothesis, we would expect this F ratio to equal approximately a value of 1.0, since if there are no mean differences, we would expect MS between to equal MS within. For our ANOVA, it is much greater than 1.0 at a value of 31.21. This F is evaluated for statistical significance on 3 and 20 degrees of freedom, yielding a p-value equal to 9.68e-08, which is extremely small, certainly smaller than conventional significance levels of 0.05 or 0.01. We reject the null hypothesis that population achievement means are equal across teachers, and conclude there is a mean difference somewhere among means, though based on the ANOVA alone, we do not know exactly where those differences may lie (we will need further analyses to tell us where, exactly).

6.4.2 Obtaining Treatment Effects

Recall that one-way ANOVA works by computing differences between sample means and the grand mean, and labeling these differences by the name of **sample** or **treatment effects**. For our data, the treatment effects are computed by the mean of the first group (71.00) minus the grand mean (79.04167), the mean of the second group (72.5) minus the grand mean, the mean of the third group (80.0) minus the grand mean, and finally, the mean of the fourth group (92.67) minus the grand mean. We can ask R to compute these values directly for us using the model.tables() function:

```
> model.tables(anova.fit)
Tables of effects

f.teach
      1       2       3       4
 -8.042  -6.542   0.958  13.625
```

The first effect, that of −8.042, is the mean difference between 71.00 and the grand mean of 79.04167, the second effect of −6.542 is the mean difference between 72.5 and the grand mean, the third effect is the mean difference between 80.0 and the grand mean, and lastly, the fourth effect is the mean difference between 92.67 and the grand mean. The sum of these effects will equal 0, as we expect them to, which recall is why we need to square treatment effects in ANOVA. However, observing the above effects gives us an immediate idea of which teach group is most different from the overall grand mean. We see that the mean achievement in the fourth group differs the most with a treatment effect equal to 13.625.

6.4.3 Plotting Results of the ANOVA

Having conducted the ANOVA, we would of course like to visualize results. A summary table of how the variance was partitioned is great, but ultimately when you present your results to an audience, whether orally through a conference presentation or through a written publication, you usually would like to give them a **picture of your findings**. As we have discussed before, nothing can be quite as powerful as a strong visualization of empirical results.

A plot of the means follows:

```
> plot.design(ac~f.teach)
```

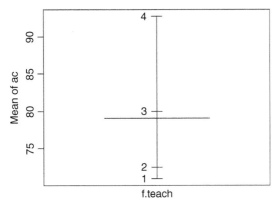

In the plot, we can see that means for teacher groups 1 and 2 are very close, while means for groups 3 and 4 are further apart. The horizontal bar situated just below the mean of group 3 is the grand mean of all the data, which recall is equal to 79.04. Hence, we can see that groups 1 and 2 are below the overall mean, while groups 3 and 4 are above it. It is always helpful to make sure what you are plotting agrees with what you already know of the data, to ensure the plot was made correctly. In our case, the plot agrees with what we already knew about the data.

6.4.4 Post Hoc Tests on the Teacher Factor

Once we have conducted the ANOVA and found evidence of mean differences through the F ratio, we often wish to then know where those mean differences lie. Recall that the F ratio does not tell us where mean differences occur; it simply tells us that somewhere among our means, there is evidence of a mean difference. To know where the difference or differences lie, we can conduct a **post hoc test**.

For our data, we will conduct the **Tukey HSD test**:

```
> anova.fit <- aov(ac~f.teach)
> TukeyHSD(anova.fit)
Tukey multiple comparisons of means
    95% family-wise confidence level

Fit: aov(formula = ac ~ f.teach)

$`f.teach`
         diff         lwr        upr       p adj
2-1   1.50000  -5.5144241   8.514424  0.9313130
3-1   9.00000   1.9855759  16.014424  0.0090868
4-1  21.66667  14.6522425  28.681091  0.0000002
3-2   7.50000   0.4855759  14.514424  0.0334428
4-2  20.16667  13.1522425  27.181091  0.0000006
4-3  12.66667   5.6522425  19.681091  0.0003278
```

We summarize the results of the post hoc:

- The first line `2-1` is the mean difference between levels 1 and 2 of the teach factor. That difference (i.e. `diff`) is equal to 1.50000. The next numbers reported are 95% confidence limits for the mean difference. The lower limit is equal to -5.5144241 while the upper limit is equal to 8.514424 (i.e. `lwr` and `upr`, respectively). That is, at a level of 95% confidence, the likely mean difference between teachers 1 and 2 is between these bounds. The final column `p adj` is the adjusted *p*-value for the mean comparison, and is equal to 0.9313130. That is, for this first comparison, there is insufficient evidence against the null to conclude a mean difference (recall that typically we would be looking for a *p*-value of 0.05 or less to conclude a difference).
- The second line `3-1` and ensuing lines are interpreted analogous to the first. For instance, the second line is the mean difference between levels 3 and 1, yielding a mean difference of 9.00000, with confidence limits 1.9855759 and 16.014424. The adjusted p-value is equal to 0.0090868 and since is less than 0.05, we deem it statistically significant.
- Overall, based on all the comparisons made, we note that we have evidence for mean differences for all pairwise comparisons except for the first. That is, all pairwise comparisons yield statistically significant differences except for `2-1`.

A great facility in R for visualizing these mean differences is again through the `plot()` function, this time plotting the results of Tukey's HSD:

```
> plot(TukeyHSD(anova.fit))
```

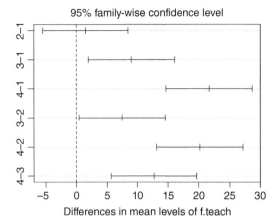

Notice in the plot that only the confidence interval 2-1 does not **exclude** a mean difference of 0, which agrees with the results of our post hoc tests above. All other confidence intervals exclude it, where 3-2 is the only other interval closest to not excluding it. This makes sense since that comparison (3-2) yielded a p-value of 0.03, quite close to the 0.05 boundary.

6.5 Alternative Way of Getting ANOVA Results via `lm()`

We have seen how we can run an ANOVA using R's `aov()` function. ANOVA, however, is part of what is known in statistics as the **general linear model**. What this means exactly is a bit beyond the scope of the current book. Interested readers should consult Fox (2016) for details and a thorough treatment of how ANOVA is a special case of the wider regression model, which in itself is a special case of the general linear model. What this means for our purposes, however, is that there is more than a single function that can generate ANOVA results in R. In what follows, we utilize the `lm()` function, which stands for "linear model" to get the ANOVA results just obtained, but whose output will differ slightly. Depending on what kind of results one prefers, using `lm()` is a legitimate option to conduct an ANOVA in R. The function `lm()` is typically used for regression-style problems, and its output will be similar to output discussed in regression chapters in this book. Hence, some of its interpretation will have to wait until we study those chapters:

```
> anova.lm <- lm(ac ~ f.teach)
> summary(anova.lm)
```

```
Call:
lm(formula = ac ~ f.teach)

Residuals:
    Min      1Q  Median      3Q     Max
-7.0000 -3.7500  0.8333  3.6250  6.0000

Coefficients:
            Estimate Std. Error t value Pr(>|t|)
(Intercept)   71.000      1.772  40.066  < 2e-16 ***
f.teach2       1.500      2.506   0.599  0.55620
f.teach3       9.000      2.506   3.591  0.00183 **
f.teach4      21.667      2.506   8.646 3.44e-08 ***
—
Signif. codes:  0 '***' 0.001 '**' 0.01 '*' 0.05 '.' 0.1 ' ' 1
Residual standard error: 4.341 on 20 degrees of freedom
Multiple R-squared:  0.824,    Adjusted R-squared:  0.7976
F-statistic: 31.21 on 3 and 20 DF,  p-value: 9.677e-08
```

We unpack as much of the above output as we can for now, and delay the rest until we study regression in a later chapter. The primary output we are interested in for now are the coefficient estimates (`Coefficients:`):

- The intercept value of 71.000 is the sample mean for the first teacher group. R is regarding this as a "baseline" group, and what follows in remaining coefficient estimates will be mean **contrasts** against this group.
- `f.teach2` is the mean comparison between the second teacher and the first. Recall that the mean of the second teacher was equal to 72.5, while the mean of the first was equal to 71.00, yielding a mean difference of 1.50. In other words, this is a mean contrast between level 2 of teach versus level 1. It is associated with a *p*-value of 0.55620, which is not statistically significant at 0.05.
- `f.teach3` is the mean comparison between the third teacher and the first, equal to 9.000. This mean contrast has an associated *p*-value equal to 0.00183 and is statistically significant.
- `f.teach4` is the mean comparison between the fourth teacher and the first, equal to 21.667, and is statistically significant ($p = 3.44e{-}08$).
- We delay our discussion of Multiple R-squared and Adjusted R-squared to the regression chapters later in the book. However, we can interpret the *F*-statistic of 31.21 on 3 and 20 degrees of freedom, with an associated *p*-value of 9.677e−08. This is simply the same *p*-value we obtained for the ANOVA using the `aov()`

function earlier. As we did then, the small *p*-value of 9.677e−08 suggests we reject the null of equal population means.

6.5.1 Contrasts in `lm()` versus Tukey's HSD

You may have noticed that these mean comparisons mimic the comparisons we obtained by using Tukey's HSD. That is, the actual mean differences are exactly the same. However, what are not the same are the *p*-values associated with each contrast. This is because in the Tukey HSD, these *p*-values were adjusted to control the type I error rate for multiple comparisons. In the ANOVA output above, *p*-values are not adjusted. This is why, for instance, the mean difference of levels 2 versus 1 is evaluated at a much more stringent level in Tukey's than it is in the above ANOVA table. Tukey's HSD did its job, and made it more difficult to reject each mean comparison.

6.6 Factorial Analysis of Variance

In the ANOVA just performed, we had only a single independent variable. Recall that achievement was hypothesized to be a function of teacher, where teacher had 4 levels. Often you may have more than a single independent variable at your disposal, and wish to include both variables simultaneously into the model. An analysis of variance having more than a single independent variable is called a **factorial analysis of variance**.

The theory behind factorial ANOVA is relatively complex, and is beyond the scope of this book to uncover it in any detail. Readers who would like a thorough treatment of the subject can turn to Denis (2016), Hays (1994), or Kirk (2012). However, one can obtain a conceptual understanding of factorial ANOVA without delving too much into detailed theory. In this chapter, we give you enough insight into the statistical method that you get on with analyzing such designs using R quickly.

6.6.1 Why Not Do Two One-Way ANOVAs?

The first question that needs to be addressed is why perform a factorial ANOVA when one can just as easily perform two one-way ANOVAs, treating each independent variable separately in each ANOVA? There are a few reasons for typically preferring the more complete factorial model over two one-way models, but by far the primary reason is that in a factorial model, the researcher will be able to model interaction terms. What is an interaction? An **interaction** in ANOVA is said to

Table 6.2 Achievement as a function of teacher and textbook.

	Teacher			
Textbook	1	2	3	4
1	70	69	85	95
1	67	68	86	94
1	65	70	85	89
2	75	76	76	94
2	76	77	75	93
2	73	75	73	91

exist when **mean differences on one factor are not consistent or the same across levels of the second factor**. In other words, when the effect of one independent variable on the dependent variable is not consistent across levels of a second independent variable. An example will help clarify the nature of interactions.

Recall the achievement data featured earlier, where achievement was hypothesized as a function of teacher. Suppose now that instead of only teacher serving as the independent variable, we hypothesize a second factor, textbook, in the same model. The data appear in Table 6.2:

A few remarks about the table:

- We see that as before, teacher is included as a factor, where there are four teachers. Effects for teachers exist in columns of the data layout.
- There is now a second factor to the design, textbook, appearing in the rows, with two levels (textbook 1 vs. textbook 2). That is, two different textbooks were used for the year.

A look at Table 6.2 reveals that we no longer have only a single hypothesis to test about teacher effects. Rather, we now have a hypothesis to test about textbook as well. That is, we now have two **main effects** in the two-way factorial ANOVA. However, we are not done. We also have a third effect, that is of the effect of the combination of **teacher and textbook**. That is, the factorial ANOVA will also help us learn whether a combination of teacher level and textbook level is associated with achievement. These combinations occur in the **cells** of the layout. To see this, consider Table 6.2 once more, but modified slightly to notate the cells as shown in Table 6.3.

Table 6.3 Achievement as a function of teacher and textbook.

	Teacher			
Textbook	**1**	**2**	**3**	**4**
1	70	69	85	95
1	67	68	86	94
1	65	70	85	89
2	75	76	76	94
2	76	77	75	93
2	73	75	73	91

When considering an interaction effect then, we are most interested in **cell means** rather than row or column means. That is, for the main effect of teacher, we are interested in column mean differences. For the main effect of textbook, we are interested in row mean differences. But for the interaction, we are interested in the **combinations of levels** of row and column factors.

As always, a plot can greatly help us make sense out of data in tables. Consider the interaction plot (Figure 6.1) of achievement by teacher and textbook. Notice in the plot that at level 1 of teach, there is a sample mean difference between achievement means (in part (b) of the plot, the differences are emphasized). Also at level 2 of the teach factor, we notice a mean difference in textbooks. For both of these teacher levels, textbook 2 is consistently above textbook 1. However, at level 3 of teacher, we see a reversal, in that textbook 1 is now much greater than textbook 2. Then, at level 4 of teach, both cell means appear to be equal. The display in Figure 6.1 (p. 168) clearly depicts an interaction (in the sample) between the factors of teacher and textbook. How do we know this? We know this because the lines are not parallel. That is, **an interaction between two factors is indicated by nonparallel lines**. In other words, **mean differences on one factor are not consistent across levels of a second factor**.

Don't FORGET! *An interaction in a two-way ANOVA is indicated by nonparallel lines. If the lines are exactly parallel (which will hardly ever be the case in practice), then no interaction between factors is said to exist. Always plot your data to be able to visualize the presence or absence of any potential interaction effects.*

6.7 Example of Factorial ANOVA

To demonstrate factorial ANOVA in R, we analyze the teacher × textbook achievement data in Table 6.2. Recall the data in R (using head() we show only the first few cases):

```
> head(achiev)
  ac teach text
1 70     1    1
2 67     1    1
3 65     1    1
4 75     1    2
5 76     1    2
```

As we did in the one-way analysis, we need to make sure both of our factors are treated as factors by R. That is, we need to convert both variables to be factors:

```
> f.teach <- factor(teach)
> f.text <- factor(text)
> f.teach
[1] 1 1 1 1 1 1 2 2 2 2 2 2 3 3 3 3 3 3 4 4 4 4 4 4
Levels: 1 2 3 4
> f.text
[1] 1 1 1 2 2 2 1 1 1 2 2 2 1 1 1 2 2 2 1 1 1 2 2 2
Levels: 1 2
```

We now fit the factorial model:

```
> fit.factorial <- aov(ac ~ f.teach + f.text + f.teach:f.
text, data = achiev)
> summary(fit.factorial)
               Df Sum Sq Mean Sq F value   Pr(>F)
f.teach         3 1764.1   588.0 180.936 1.49e-12 ***
f.text          1    5.0     5.0   1.551    0.231
f.teach:f.text  3  319.8   106.6  32.799 4.57e-07 ***
Residuals      16   52.0     3.3
---
Signif. codes: 0 '***' 0.001 '**' 0.01 '*' 0.05 '.' 0.1 ' ' 1
```

We summarize the output:

- The effect for f.teach has 3 degrees of freedom, computed as the number of levels of the factor minus 1. Since there are 4 teach levels, degrees of freedom are equal to $4 - 1 = 3$. It has associated with it a sum of squares equal to 1764.1, which when we divide by degrees of freedom, yields a mean square of 588.0.

We will use this mean square shortly in the computation of the *F*-ratio. For effect size, **partial Eta-squared**, η_p^2, can be used in conjunction or instead of ordinary Eta-squared. Partial Eta-squared, as the name suggests, "partials" out other factors from the effect size estimate. For teach, it is computed as SS teach/(SS teach + SS residuals) = 1764.1/(1764.1 + 52) = 1764.1/1816.1 = 0.97. That is, approximately 97% of the variance in achievement is accounted for by teach, partialling out the effect of text (notice text did not make its way into the denominator).

- The effect for f.text has 1 degree of freedom, computed as the number of levels of the factor minus 1. Since there are 2 textbook levels, degrees of freedom are equal to $2 - 1 = 1$. It has associated with it a sum of squares equal to 5.0, which when we divide by degrees of freedom, yields a mean square of 5.0. We will use this mean square shortly in the computation of the *F*-ratio.
- The test for an interaction between teacher and textbook is indicated by f.teach:f.text. It has associated with it 3 degrees of freedom, computed as the degrees of freedom for teach multiplied by the degrees of freedom for text (i.e. 3×1). Its sum of squares is equal to 319.8, which when divided by degrees of freedom, yields a mean square of 106.6. As with the main effect terms, this mean square will likewise be used to obtain the corresponding *F*-ratio.
- The Residuals term is the "left over" of the model, that which is unexplained by prior terms entered into the model. It has associated with it 16 degrees of freedom, and has a sum of squares value of 52.0. Its mean square is equal to $52.0/16 = 3.3$.
- An *F*-ratio is computed for each term in the analysis. For each term tested, we compare the corresponding mean square to residual mean square. For teach, the computation is $588.0/3.3 = 180.936$. For text, the computation is $5.0/3.3 = 1.551$. For the teach by text interaction, the computation is $106.8/3.3 = 32.799$. All *F* statistics are evaluated for statistical significance. The effect for teach is statistically significant ($p = 1.49e-12$), while the effect for text is not ($p = 0.231$). The effect for interaction is statistically significant ($p = 4.57e-07$).

We can inspect the cell means more closely in R using the **phia** package:

```
> library(phia)
> (fit.means <- interactionMeans(fit.factorial))
  f.teach f.text adjusted mean std. error
1       1      1      67.33333   1.040833
2       2      1      69.00000   1.040833
3       3      1      85.33333   1.040833
4       4      1      92.66667   1.040833
5       1      2      74.66667   1.040833
6       2      2      76.00000   1.040833
7       3      2      74.66667   1.040833
8       4      2      92.66667   1.040833
```

(a)

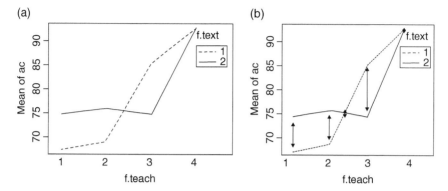

(b)

Figure 6.1 (a) Cell means for teacher * textbook on achievement. (b) Distances between cell means as depicted by two-headed arrows (where f.text is the factor name for textbook and f.teach is the factor name for teacher).

The above displays the cell means for each combination of teacher and text that corresponds to Figure 6.1. For example, in level 1 for both teacher and text, the mean is equal to 67.33. R calls it the "adjusted mean," but it's simply the mean of achievement for the given cell. Try to see if you can match up these cell means with the plot in Figure 6.1. We summarize the cell means plot in Table 6.4.

To obtain a plot of the interaction effect that we featured earlier, we use inter-action.plot() in R, which is part of the base package:

```
> interaction.plot(f.teach, f.text, ac)
```

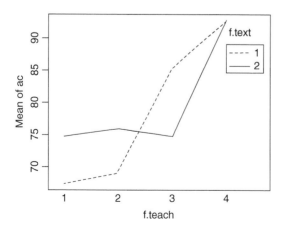

Table 6.4 Achievement cell means teacher * textbook.

Textbook	Teacher				Row means
	1	**2**	**3**	**4**	
1	$\bar{y}_{jk} = \bar{y}_{11} = 67.33$	$\bar{y}_{jk} = \bar{y}_{12} = 69.00$	$\bar{y}_{jk} = \bar{y}_{13} = 85.33$	$\bar{y}_{jk} = \bar{y}_{14} = 92.67$	$\bar{y}_{j.} = \bar{y}_{1.} = 78.58$
2	$\bar{y}_{jk} = \bar{y}_{21} = 74.67$	$\bar{y}_{jk} = \bar{y}_{22} = 76.00$	$\bar{y}_{jk} = \bar{y}_{23} = 74.67$	$\bar{y}_{jk} = \bar{y}_{24} = 92.67$	$\bar{y}_{j.} = \bar{y}_{2.} = 79.50$
Column means	$\bar{y}_{.k} = \bar{y}_{.1} = 71.00$	$\bar{y}_{.k} = \bar{y}_{.2} = 72.5$	$\bar{y}_{.k} = \bar{y}_{.3} = 80.0$	$\bar{y}_{.k} = \bar{y}_{.4} = 92.67$	$\bar{y}_{..} = 79.04$

As we did when plotting the main effect for teach, we can generate a plot that simultaneously features the two main effects for teach and text. We can use `plot.design()` for this:

```
> plot.design(ac~f.teach + f.text + f.teach:f.text, data =
achiev)
```

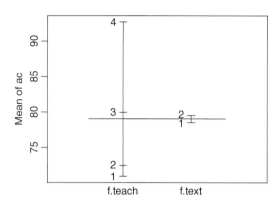

Factors

The teach mean differences are as before, but now we see the text differences. As we can see from the plot, there is very little evidence of textbook differences. We can obtain plots of the relevant row means for teacher using the `interactions-Means()` function:

```
> interactionMeans(fit.factorial, factors="f.teach")
  f.teach adjusted mean std. error
1       1      71.00000  0.7359801
2       2      72.50000  0.7359801
3       3      80.00000  0.7359801
4       4      92.66667  0.7359801
```

Let's now perform a Bonferroni test on teach:

```
> pairwise.t.test(ac, f.teach, p.adj = "bonf")

Pairwise comparisons using t tests with pooled SD

data:  ac and f.teach

  1        2        3
2 1.00000  -        -
3 0.01095  0.04316  -
4 2.1e-07  6.4e-07  0.00036
P value adjustment method: bonferroni
```

Above are reported *p*-values for each comparison. Be careful to note that these numbers are not actual mean differences. They are *p*-values. That is, in the cell in the top left of the table, for instance, where 1.00000 appears, this is the *p*-value associated with comparing the first mean of teach to the second mean. Since those means are very close (71 vs. 72.5), the Bonferroni *p*-value for the test is rounded up to 1.0 (it is not exactly equal to 1, but is very close to it). We see that the comparison of means 2 and 4 has a *p*-value equal to 6.4e−07, which is extremely small. This is not surprising, since that mean difference is quite large (72.5 vs. 92.67).

We will now obtain the row means for the textbook effect:

```
> library(phia)
> interactionMeans(fit.factorial, factors="f.text")
  f.text adjusted mean std. error
1      1      78.58333  0.5204165
2      2      79.50000  0.5204165
```

Again, we see that the means across textbook are very close. Remember that since we have only two means, performing any kind of post hoc comparison on these two means doesn't make any sense, since, of course, there is no significance level to adjust when you only have two means.

6.7.1 Graphing Main Effects and Interaction in the Same Plot

A very useful utility in the phia package is the ability to plot interactions that show the means at each level of each factor. We can accomplish this using the plot() feature on the object fit.means that we obtained earlier:

```
> library(phia)
> plot(fit.means)
```

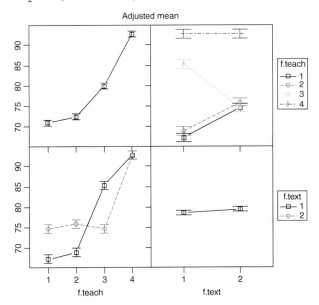

In the plot, the main effect of teach is shown in the upper left quadrant, while the main effect of text is given in the lower right. Interaction plots are given in the off-diagonals. The plot is both color-coded as well as coded with symbols (which works nicely since the figure is printed in black and white in this book).

6.8 Should Main Effects Be Interpreted in the Presence of Interaction?

Having discussed the concept of an interaction, and having seen it in motion in our achievement example, a question that naturally arises is whether main effects should be interpreted in the presence of an interaction effect. That, if an interaction is found in an ANOVA, are main effects interpretable?

While there are differing opinions regarding how to treat main effects in the presence of interaction, it is a fact that the interaction term will always contain the more **complete story** that the data are revealing. For example, for our data, we could, if we really wanted to, conclude there is a teacher effect without mentioning the interaction term, since the p-value for teach does come out to be statistically significant. However, reporting an effect for teach does not tell the most complete story. Why not? Because though achievement means increase across teach, that difference is mitigated (or "moderated" if you prefer) by what level of textbook is considered.

For instance, in looking at the interaction plot, we see that achievement means increase rather dramatically after teachers 1 and 2. However, notice that at teacher 3, for textbook 2, the mean has not increased from what it was at teacher 2. The overall mean for teacher 3 increased because of the impact of the first textbook. Hence, if we concluded, "the mean for achievement rises quickly from teach = 2 to teach = 3," though we would technically be correct, the conclusion would be incomplete and a bit misleading since it does not reveal that the reason for this increase is primarily due to the influence of textbook 1, not textbook 2. It is for reasons such as this that while reporting and interpreting main effects in the presence of interaction is permissible at a scientific level, it is generally not advisable in most cases and may masquerade the data. The general rule is that if there is an interaction in your data, it is the **interaction that should be interpreted first**, and should guide your overall substantive conclusions.

Indeed, it does not take long to realize how easily it may be for marketers and advertisers with knowledge of statistics to "massage" conclusions to their liking. Suppose a test for a medication is performed, and overall, there is a main effect for it. That is, overall, taking the medication reduces symptoms. However, if there is also an interaction effect found such that the reduction in symptoms only occurs for persons aged 40 or less, then concluding "an effect for medication" would be misleading not to mention unethical. An elderly person taking the medication may not stand to benefit at all from it, even if the advertisement on the package reporting that it reduces symptoms is, technically speaking, a true statement. The failure to report interactions is hence a masterful way to mislead with statistics, and readers of research reports need to be acutely aware of this potential.

Although technically main effects can be interpreted in the presence of an interaction, in most cases it is the interaction term that will provide the most complete portrayal of the scientific result. At minimum, reporting only main effects and purposely suppressing the reporting of interaction effects is not only scientifically a poor decision, but is also ethically unwise.

6.9 Simple Main Effects

Usually, after spotting an interaction effect, a researcher would like to conduct what are known as **simple main effects**. These essentially "break down" the interaction effect and give us more insight about what generated it in the first place. For our data, having obtained evidence for a teacher by textbook interaction, a researcher may then be interested in evaluating mean differences of **textbook at each level of teacher**. These are called simple main effects because they are the effect of one factor at a given level of the other, effectively breaking down the interaction into constituent parts so we can inspect it more closely.

In R, we can conduct simple main effects again using the phia package. As an example of a simple effects analysis, let's analyze the effect of textbook at each level of teacher:

```
> library(phia)
> testInteractions(fit.factorial, fixed="f.teach",
across="f.text")
F Test:
P-value adjustment method: holm
          Value Df Sum of Sq      F    Pr(>F)
1       -7.3333  1    80.667 24.820 0.0004071 ***
2       -7.0000  1    73.500 22.615 0.0004299 ***
3       10.6667  1   170.667 52.513 7.809e-06 ***
4        0.0000  1     0.000  0.000 1.0000000
Residuals       16    52.000
—
Signif. codes: 0 '***' 0.001 '**' 0.01 '*' 0.05 '.' 0.1 ' ' 1
```

We unpack the above output:

- By default, the phia package conducts the **Holm test** for these simple effects. The Holm test is a multistage post hoc test, somewhat like the Bonferroni; however, it adjusts for the number of comparisons contingent on the number of null hypotheses being tested. For more details on this test, see Howell (2002, pp. 386–387). Regardless, the point is that some adjustment has been made for the fact that we are conducting more than a single simple effect on the same data. If it were not the Holm test, then some other adjustment would likely be advised, such as the aforementioned Bonferroni (which divides the significance level equally across the number of comparisons).
- The first row is the textbook mean difference at the first teacher. This mean difference is equal to -7.3333. Note that the **sign** of the difference is immaterial here from a statistical point of view (though of course it may matter to your scientific hypothesis). What matters is the magnitude. That is, we could have just as easily subtracted one textbook level from the other. The order in which we are computing the mean difference is immaterial. The p-value associated with this difference is 0.0004071, which is statistically significant at $p < 0.001$. That is, we have evidence of a mean difference of textbook at the first teacher.
- The second and remaining rows are interpreted analogously. For teacher 2, we again have a statistically significant difference in means yielding a sample difference of -7.0000, and also statistically significant ($p = 0.0004299$).
- We see that the only mean comparison not statistically significant is at teacher 4, where the mean sample difference is equal to 0.0000, yielding a p-value of 1.0.

We could also, of course, evaluate teacher differences at each textbook:

```
> testInteractions(fit.factorial, fixed="f.text", across="f.teach")
F Test:
P-value adjustment method: holm
          f.teach1 f.teach2 f.teach3 Df  Sum of Sq      F    Pr(>F)
1          -25.333  -23.667  -7.3333  3     1386.9 142.248 1.911e-11
2          -18.000  -16.667 -18.0000  3      697.0  71.487 1.745e-09
Residuals                             16      52.0

1           ***
2           ***
Residuals
- - -
```

```
Signif. codes: 0 '***' 0.001 '**' 0.01 '*' 0.05 '.' 0.1 ' ' 1
```

We see that at each textbook, there exists mean teacher differences; however, the output requires a bit of explanation. The baseline or reference category here is the fourth teach group. Hence, the first row (1) in the above output, that of

```
1              -25.333   -23.667   -7.3333
```

features the mean differences of teacher levels 1, 2, and 3 against level 4, respectively. That is, the first number of -25.333 is the mean difference between the first level of teach and that of the fourth at textbook 1. The next difference of -23.667 is the mean difference between the second level of teach and the fourth, again, at textbook 1.

The second line is interpreted analogously to the first, only that now, we are featuring mean teacher differences at textbook 2:

```
2              -18.000   -16.667  -18.0000
```

The first number of -18.000 is the mean difference between the first teacher and the fourth at textbook 2. The second difference of -16.667 is the mean difference between the second teacher and that of the fourth, again at textbook 2. And so on for the remainder of the differences.

6.10   Random Effects ANOVA and Mixed Models

So far in this chapter, we have surveyed the one-way ANOVA as well as the two-way ANOVA, which were special cases of the wider and more general factorial ANOVA model. However, in all of our dealings thus far, we have assumed that effects in all models are **fixed effects**. This is important, since how we treat factors in ANOVA depends very much on whether we consider them fixed or random. A factor is a fixed effect if the levels we are using in the experiment or study are the only ones we are interested in generalizing our results to. In our example featured in ANOVA thus far, both teacher and textbook were assumed to be fixed effects, in that we were only interested in differences (and interaction) between the particular levels chosen for the experiment and did not wish to generalize to other levels. That is, we were only presumably interested in teachers 1 through 4, those **exact particular teachers**, and were not interested in generalizing our results to other teachers that could have theoretically been selected for the study. Hence, determining mean differences in this regard between teachers made perfect sense, because of our interest in specific teachers. Analogously, we were only interested in textbooks 1 and 2, and not other textbooks that we could have potentially selected. This is what made textbook a fixed effect.

A factor is considered **random** in ANOVA if we wish to generalize to levels of the factor of which the levels appearing in our experiment or study are merely a **sample**. That is, a factor is random if the levels are chosen in a random fashion out of a wider population of levels that could have theoretically been selected for the given experiment. For instance, instead of only being interested in teachers 1 through 4, perhaps the investigator is interested in the population of teachers of which the 4 selected constitute a random sample. Hence, mean comparisons among the 4 teachers selected would no longer make sense, but rather a measure of how much **variance** is attributable to the teacher factor would be of more interest. We expand on this explanation in what follows, unpacking why an investigator might want to consider teacher random in our featured experiment.

6.10.1 A Rationale for Random Factors

The reason why a researcher may wish to perform a random effects ANOVA over the more traditional fixed effects ANOVA is to make a wider generalization of variance accounted for by the factor to more levels than what appeared in the experiment or study. For instance, consider again our example of achievement as a function of teacher. In the fixed model, we could only make conclusions about the 4 teachers selected, and thus were interested in mean differences between these particular teachers. However, perhaps the researcher might be more interested in drawing conclusions about teachers in general. Hence, to conduct her ANOVA, she may wish to draw a **random sample of teachers from a population of teachers**, which will allow her to draw a conclusion about the population of teachers in general rather than just the 4 teachers she happened to select when sampling. This could provide the investigator with a more powerful research statement, yet at the expense of not being able to make precise statements about mean differences among the teachers that happened to show up in the experiment. Let's compare the fixed to the random model in the following conclusions, given evidence for an effect of teacher (i.e. suppose teacher comes out to be statistically significant):

- **Fixed Factor** – there is evidence of a mean difference on achievement somewhere among teachers.
- **Random Factor** – teachers, either those sampled for the given experiment or teachers in general, are associated with variance explained in achievement.

Having briefly reviewed the differences between fixed and random factors, let's now survey an example in R featuring the case of achievement as a function of teacher. A word of warning – random effects models (and mixed models, as we shall soon see) are a bit of different "animals" compared to fixed effects, in that they have more options for how to estimate parameters and their output can appear much different than in fixed effects. So, what follows is really only the tip of the iceberg when it comes to random and mixed models. We make necessary referrals to other sources where required. Entire courses and careers are spent on random effects and mixed models.

6.10.2 One-Way Random Effects ANOVA in R

To demonstrate a one-way random effects analysis of variance (i.e. a random effects ANOVA having only a single independent variable), we re-analyze the achievement data, this time specifying teacher as **random**. For our example, it is the only factor in the experiment. We can fit a one-way random effects model using R's lme4 () package. As mentioned, fitting a random effects model is a bit more complex than fixed, and the coding of the model requires a bit more explanation, so we take it step by step:

```
> library(lme4)
> fit.random <- lmer(ac ~ 1 + (1|f.teach), achiev, REML = FALSE)
```

In the above call statement, fit.random is of course the name we are giving to the object. The lmer() function is similar in spirit to the simpler lm() model for an ordinary linear model (recall we used this function for running our ANOVA model earlier). Next, we write the dependent variable ac as a function of an intercept value, denoted by 1 plus the statement (1|f.teach) which specifies teach as a random factor. Next, achiev is simply the name of the data frame as in previous models. The final statement REML = FALSE is requesting maximum likelihood (ML) estimation instead of REML, which is **restricted maximum likelihood**. Turning REML "off" generates **ML** estimates.

We now run the model and obtain a summary:

```
> summary(fit.random)
Linear mixed model fit by maximum likelihood  ['lmerMod']
Formula: ac ~ 1 + (1 | f.teach)
   Data: achiev

     AIC      BIC   logLik deviance df.resid
   157.2    160.7    -75.6    151.2       21

Scaled residuals:
    Min      1Q  Median      3Q     Max
-1.6032 -0.8810  0.2963  0.7671  1.3917
```

```
Random effects:
 Groups     Name            Variance  Std.Dev.
 f.teach   (Intercept)      70.37     8.389
 Residual                   18.84     4.341
Number of obs: 24, groups:  f.teach, 4

Fixed effects:
             Estimate Std. Error t value
(Intercept)   79.042      4.287    18.44
```

The output from a random-effects model, as we have warned, looks somewhat different from the traditional ANOVA output, and requires some explanation. We unpack the most important of the above output items:

- The **AIC** statistic is a measure of model fit. For our data, AIC is equal to 157.2. Smaller values than not are better for an AIC statistic. For instance, if we included a second factor into our model, we may be interested to learn whether AIC drops by a fair amount from the value we obtained above of 157.2. AIC essentially "punishes" one for including predictors (or factors) that are not worthwhile statistically, in the sense that they do not sufficiently contribute to model fit. We could compare a few models to our tested model to see which generates the **lowest AIC**.
- BIC yields similar information as AIC, and has an analogous role in helping determine the "best" model based on "smaller is better" values. However, BIC tends to punish the user a bit more heavily for including variables in the model that do not contribute substantially (i.e., it invokes a larger penalty than AIC). For our data, BIC is equal to 160.7.
- We next interpret the random effects, of which there are two: f.teach and Residual. Of course, our substantive interests lie with f.teach, and we see it boasts a **variance component** of 70.37, which is the square of the standard deviation next to it of 8.389.
- The **residual variance** component is equal to 18.84, the square of 4.341.
- R next lists the fixed effects in the model. The only fixed effect in this case is the intercept term, which yields a grand mean of 79.042.

Because teacher is a random factor in this model, we are not so much interested in precise mean teacher differences as we are about how much variance is accounted for by the teacher factor, either by those levels of teacher that happened to be sampled in the given experiment or those in the population from which these teachers were drawn. To obtain the proportion of variance explained, we can compute what is known as the **intraclass correlation**:

$$\frac{\sigma_A^2}{\sigma_Y^2} = \frac{70.36}{70.36 + 18.84} = 0.79$$

We interpret the intraclass correlation as follows: approximately 79% of the variance in achievement scores is attributable to teacher differences.

Recall that we had estimated the above model by turning off REML estimation, thus requesting ML. To fit the same model via REML, we simply code the following:

```
> fit.random.reml <- lmer(ac ~ 1 + (1|f.teach), achiev, REML = TRUE)
> summary(fit.random.reml)
Linear mixed model fit by REML ['lmerMod']
Formula: ac ~ 1 + (1 | f.teach)
   Data: achiev

REML criterion at convergence: 146.3

Scaled residuals:
    Min      1Q  Median      3Q     Max
-1.6056 -0.8696  0.2894  0.7841  1.3893

Random effects:
 Groups   Name        Variance Std.Dev.
 f.teach  (Intercept) 94.87    9.740
 Residual             18.84    4.341
Number of obs: 24, groups:  f.teach, 4

Fixed effects:
            Estimate Std. Error t value
(Intercept)    79.04       4.95   15.97
```

In the above, we did not have to explicitly state REML = TRUE. We could have just as easily left the statement out altogether, and R would have understood this to imply that REML estimation was requested. That is, we could have simply coded:

```
> fit.random.reml <- lmer(ac ~ 1 + (1|f.teach), achiev).
```

The model fit by REML, not surprisingly, is very similar to that fit by ML; however, the proportion of variance explained by teacher is a bit higher. In this model, it is equal to $94.87/(94.87 + 18.84) = 0.83$. That is, approximately 83% of the variance in achievement is accounted for by the teacher factor using REML estimation. This is an increase of about 4% over that obtained using ML estimation.

The differences between ML and REML estimation are far beyond the scope of this book. Interested readers who want an in depth discussion of these estimation methods are encouraged to consult Searle et al. (2006). If we had to give an overall recommendation, it would be to choose REML estimation over ML estimation for most models, since this is what Searle et al., also generally recommend.

6.11   Mixed Models

Recall that the fixed-effects model has only fixed factors in it, while the purely ran-dom-effects model has only random factors. If a model has one or more fixed fac-tors and one or more random factors, it is referred to as a **mixed model**. To fit a mixed model, we can use the **nlme** package in R. For this example, we will des-ignate textbook as fixed, and teacher as random:

```
> library(nlme)
> mixed.model <- lme(ac ~ f.text, data = achiev, random =
~1 | f.teach)
> summary(mixed.model)

Linear mixed-effects model fit by REML
 Data: achiev
       AIC      BIC    logLik
  151.0375 155.4017 -71.51875

Random effects:
 Formula: ~1 | f.teach
        (Intercept) Residual
StdDev:    9.733736 4.423571

Fixed effects: ac ~ f.text
               Value Std.Error DF   t-value p-value
(Intercept) 78.58333  5.031607 19 15.617940  0.0000
f.text2      0.91667  1.805915 19  0.507591  0.6176
 Correlation:
        (Intr)
f.text2 -0.179

Standardized Within-Group Residuals:
       Min         Q1        Med        Q3        Max
-1.6788343 -0.7678930  0.1811275 0.7352230 1.4671909

Number of Observations: 24
Number of Groups: 4
```

As we can see, the call statement for the model is very similar to that of a random-effects model, only that now, ac ~ f.text specifies textbook as a fixed factor, while random = ~1 | f.teach designates teach as a random factor (more concisely, it gives teacher a random intercept, though understanding this interpretation is not imperative, it is enough to know that the statement is designating teacher as a random effect). The output f.text2 is a mean contrast between the first and second textbooks. That is, the number 0.91667 is an estimate of the mean difference between textbooks. Since textbook is a fixed factor, estimating mean differences is of interest. The variance component for f.teach is equal to 9.73 squared, which is 94.67. The square of the residual term is equal to 19.57, and so the proportion of variance explained by f.teach is equal to 94.67/ (94.67 + 19.57) = 0.83. That is, as we saw earlier, approximately 83% of the variance in achievement is accounted for by teachers, either those sampled for the given study or those in the population from which the given sample was drawn.

6.12 Repeated-Measures Models

In models we have surveyed up to now, each group or cell (in the case of factorial models) had different subjects or participants. That is, a participant in group 1, for instance, was not also simultaneously in group 2. However, many times in applied social and natural sciences, researchers are interested in tracking participants across time, and seek to measure them repeatedly. In such models, the grouping factor actually contains the same subject tested more than once. These models typically go by the name of **repeated-measures ANOVA**, or **longitudinal models** more generally. As an example of a design that would call for a repeated-measures

Table 6.5 Learning as a function of trial (hypothetical data).

Rat	Trial			Rat means
	1	2	3	
1	10.0	8.2	5.3	7.83
2	12.1	11.2	9.1	10.80
3	9.2	8.1	4.6	7.30
4	11.6	10.5	8.1	10.07
5	8.3	7.6	5.5	7.13
6	10.5	9.5	8.1	9.37
Trial means	$M = 10.28$	$M = 9.18$	$M = 6.78$	

analysis, consider the layout in Table 6.5, where rats were measured on the elapsed time required to press a lever in an operant conditioning chamber.

A few things to note about the table:

- These data are of a different kind than in between-subjects designs, since each participant (rat, in this case) is measured repeatedly. For example, the first rat obtained scores of 10.0, 8.2, and 5.3, for a combined row mean of 7.83. Notice that these scores are not on different rats, but rather on the **same rat**.
- As one might expect for the learning task, the mean number of errors decreases across trials. That is, the mean for trial 1 is 10.28, the mean for trial 2 is 9.18, and the mean for trial 3 is 6.78. The data seem to suggest that rats, on average, are learning the task across trials.

In our repeated-measures data then, we can identify two sources of variability, that variability occurring **between trials**, and that variability occurring **within rat**, or more generally, **within-subject** (however, in this case since we are working with rats, we are calling it "within rat"). So what makes repeated-measures models so much different than ordinary between-subjects models? The key difference is that **we can no longer expect that scores between trials are uncorrelated** as we would in a purely between-subjects design. Recall that in a between-subjects model, the expectation was that scores in different groups of the independent variable were not related. For instance, if a student was assigned teacher 1, we had no reason to believe that student's score would be related to another student assigned teacher 2. The situation however is different in repeated-measures models. We now expect a **correlation** between levels of the independent variable. That is, for our data, we now expect trials to be related. Repeated-measures models take this correlation into account when generating ensuing F-ratios.

Let's get the data into R, and begin looking at it more closely:

```
> learn <- read.table("learning.txt", header = T)
> learn
  rat trial time
1   1     1 10.0
2   1     2  8.2
3   1     3  5.3
4   2     1 12.1
5   2     2 11.2
6   2     3  9.1
7   3     1  9.2
8   3     2  8.1
9   3     3  4.6
```

```
10    4        1  11.6
11    4        2  10.5
12    4        3   8.1
13    5        1   8.3
14    5        2   7.6
15    5        3   5.5
16    6        1  10.5
17    6        2   9.5
18    6        3   8.1
```

A plot of the data will reveal the trend we noticed by inspecting the data layout:

```
< attach(learn)
> library(ggplot2)
> qplot(trial, time)
```

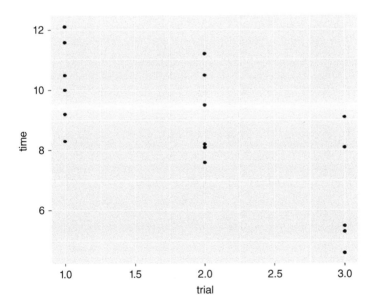

The plot depicts the trend we observed in the data, that the number of errors decreases across trials.

We run the repeated measures in R:

```
> f.rat <- factor(rat)
> f.trial <- factor(trial)
> rm.model <- aov(time ~ f.trial + Error(f.rat/f.trial), data =
learn)
```

```
> summary(rm.model)

Error: f.rat
          Df Sum Sq Mean Sq F value Pr(>F)
Residuals  5  35.62   7.124

Error: f.rat:f.trial
          Df Sum Sq Mean Sq F value   Pr(>F)
f.trial    2  38.44  19.220   72.62 1.11e-06 ***
Residuals 10   2.65   0.265
---
Signif. codes:  0 '***' 0.001 '**' 0.01 '*' 0.05 '.' 0.1 ' ' 1
```

We note there is a statistically significant effect of f.trial ($p = 1.11e{-}06$). That is, we have evidence of a mean difference in learning across trials. We would of course next be interested in where this difference or differences may lie, and for that, we can turn to post hoc tests. The packages **emmeans** and **multcomp** can conduct a wide variety of post hocs for ANOVA-type models.

Suppose instead of only having the single factor of trial, we also assigned rats to different treatments, making it a two-way repeated-measures model, with one factor within (trial) and one factor between (treatment). The updated layout is in Table 6.6:

Table 6.6 Learning as a function of trial and treatment (hypothetical data).

Treatment	Rat	Trial 1	Trial 2	Trial 3	Rat means
Yes	1	10.0	8.2	5.3	7.83
No	2	12.1	11.2	9.1	10.80
Yes	3	9.2	8.1	4.6	7.30
No	4	11.6	10.5	8.1	10.07
Yes	5	8.3	7.6	5.5	7.13
No	6	10.5	9.5	8.1	9.37
	Trial means	$M = 10.28$	$M = 9.18$	$M = 6.78$	

To run this model, we update our data to include the treatment factor:

```
> learn
   rat trial time treat
1    1      1 10.0     1
2    1      2  8.2     1
3    1      3  5.3     1
4    2      1 12.1     2
5    2      2 11.2     2
6    2      3  9.1     2
7    3      1  9.2     1
8    3      2  8.1     1
9    3      3  4.6     1
10   4      1 11.6     2
11   4      2 10.5     2
12   4      3  8.1     2
13   5      1  8.3     1
14   5      2  7.6     1
15   5      3  5.5     1
16   6      1 10.5     2
17   6      2  9.5     2
18   6      3  8.1     2
```

Next, we designate treat as a factor:

```
> attach(learn)
> f.treat <- factor(treat)
> f.treat
 [1] 1 1 1 2 2 2 1 1 1 2 2 2 1 1 1 2 2 2
Levels: 1 2
```

We are now ready to run the model:

```
> rat.two.way <- aov(time~f.trial*f.treat + Error(f.rat/f.
trial), data = learn)
> summary(rat.two.way)

Error: f.rat
           Df Sum Sq Mean Sq F value   Pr(>F)
f.treat     1  31.73   31.73   32.68  0.00463 **
Residuals   4   3.88    0.97
---
```

```
Signif. codes:  0 '***' 0.001 '**' 0.01 '*' 0.05 '.' 0.1 ' ' 1
Error: f.rat:f.trial
                  Df Sum Sq Mean Sq F value    Pr(>F)
f.trial            2  38.44  19.220  91.403 3.09e-06 ***
f.trial:f.treat    2   0.96   0.482   2.293   0.163
Residuals          8   1.68   0.210
---
Signif. codes:  0 '***' 0.001 '**' 0.01 '*' 0.05 '.' 0.1 ' ' 1
```

We find evidence for a trial ($p = 3.09e{-}06$) and treat effect (0.00463); however, not for an interaction effect ($p = 0.163$).

We have only literally scratched the surface here with regard to repeated-measures models to demonstrate their use in R. There are entire books devoted to the subject, as well as to longitudinal models in general. Careers too are built on studying and developing longitudinal models. One of the better books on repeated measures and ANOVA models in general is Kirk (2012), and should be consulted especially if you are an experimental scientist who regularly runs these types of models. Fitzmaurice (2011) is an entire book devoted to the subject. Repeated-measures models also involve additional assumptions (such as sphericity) that we have not discussed here, but are discussed at length in sources such as Kirk. Hays (1994) also discusses many of the assumptions underlying these models as a special case of the randomized block design.

Exercises

1   Discuss how the ANOVA model can be conceptualized as both an extension of univariate *t*-tests, as well as a special case of the wider regression model. You may wish to study regression in the following chapter before attempting the second component of this question.

2   When $n_1 = n_2$ in the *t*-test, is pooling variances required? Why or why not?

3   Explain the overall goal of ANOVA by referring to the measures of between versus within variance.

4   Your colleague obtained sample means of 32, 31, and 30 for a three-group problem in ANOVA. He claims, "Since the means are different, this is cause to reject the null hypothesis." Is your colleague correct? Why or why not? Explain.

5 Explain the concept of an interaction in ANOVA using a hypothetical example of your choosing.

6 Consider the iris data:

```
> head(iris)
  Sepal.Length Sepal.Width Petal.Length Petal.Width
Species
1          5.1         3.5          1.4         0.2 setosa
2          4.9         3.0          1.4         0.2 setosa
3          4.7         3.2          1.3         0.2 setosa
4          4.6         3.1          1.5         0.2 setosa
5          5.0         3.6          1.4         0.2 setosa
6          5.4         3.9          1.7         0.4 setosa
```

Conduct a fixed-effects one-way ANOVA where Sepal.Length is hypothesized as a function of Species. To run the ANOVA, use both the aov() and lm() functions, and compare your results.

7 Conduct a Tukey test on the ANOVA performed in #6, and briefly summarize your findings.

8 Conduct a factorial ANOVA on the achievement data with teacher and text as independent variables, but this time, operationalize the teach factor into two levels only: where teachers 1–2 are level 1, and teachers 3–4 are level 2. Summarize your findings from the ANOVA, then obtain a plot of cell means using the **phia** package. Finally, conduct a simple effects analysis analyzing teacher differences at each level of text. Comment on your findings.

9 For #8, did it make most sense to regard teacher and text as fixed, or random effects? Why?

10 In #6, you conducted a one-way ANOVA, assuming a fixed model. Now, again perform a one-way ANOVA, but this time a one-way random-effects ANOVA, where the Species factor is random. Interpret findings, paying particularly close attention to the interpretation of the Species factor as random. How is the interpretation different than when Species was fixed? Explain.

7

Simple and Multiple Linear Regression

LEARNING OBJECTIVES

- Understand how linear regression works, and the differences between simple linear regression and multiple regression.
- Understand what least-squares accomplishes, and why it guarantees to minimize squared error, not necessarily make it small for any given data set.
- How to compute the intercept and slope in a simple linear regression, and how to interpret coefficients.
- Understand the difference between R^2 and adjusted R^2, and why the latter is often preferred over the former.
- Why multiple regression is necessarily more complex, both statistically and scientifically than simple linear regression, and understand why all models are context-dependent.
- Understand and be able to use model-building strategies such as forward and stepwise regression, and understand drawbacks and caveats regarding these approaches and why others may be preferable.

Regression analysis is at the very heart of applied statistics. The goal of regression analysis is to make predictions on a continuous response variable based on one or more predictor variables. If there is only a single predictor variable in the model, it is called **simple linear regression**. If there is more than a single predictor, it is referred to as **multiple linear regression**. Multiple linear regression seeks to model the simultaneous prediction of several predictor variables on a single response variable. In the case where there are several predictors and several responses, this is an example of **multivariate multiple regression**, a topic beyond the scope of this book.

When you think about it, virtually all science is about making predictions. Aside from seeking to determine "true causes," which according to some quantum physicists may be more complicated than traditional theories of causation would

Univariate, Bivariate, and Multivariate Statistics Using R: Quantitative Tools for Data Analysis and Data Science, First Edition. Daniel J. Denis.
© 2020 John Wiley & Sons, Inc. Published 2020 by John Wiley & Sons, Inc.

suggest, if a scientist can predict an outcome, that scientist is usually quite credible. Examples of prediction might include the following:

- Predicting symptoms to subside after taking a headache medication
- Predicting probability of cancer based on genetic history
- Predicting whether a vehicle will break down based on maintenance history
- Predicting suicide probability based on suicidal ideation (i.e. one's thinking of suicide)

We can say then, that in the big picture of things, science, and much of daily life for that matter, is about **prediction**. Regression analysis is the "king" of predictive methods in statistics, and a whole variety of statistical methods actually derive from it, or can be considered special cases of the method. The ANOVA models we featured in the previous chapter can actually be considered special cases of regression. Regression analysis has its origins much before ANOVA beginning in the late 1800s, though least-squares was first discovered at the turn of the nineteenth century with the work of Gauss and Legendre (Stigler, 1986). ANOVA came on the scene later in early 1900s, and featured a way for experimental scientists to use statistical methods to help interpret data from experiments.

7.1 Simple Linear Regression

We begin with a discussion of simple linear regression, then extend our discussion into multiple regression a bit later. We start with an applied example to demonstrate the procedure before delving into a bit more of the theory to explain what the output means, which will include a discussion of least-squares. Consider hypothetical IQ data:

```
> iq.data <- read.table("iq.data.txt", header = T)
> some(iq.data)
   verbal quant analytic group
1      56    56       59     0
2      59    42       54     0
3      62    43       52     0
4      74    35       46     0
5      63    39       49     0
6      68    50       36     0
```

For now, suppose an investigator would like to predict one's verbal ability based on knowledge of her quantitative ability. Let's first produce a scatterplot to see if there might be a relation:

```
> attach(iq.data)
> scatterplot(verbal ~ quant)
```

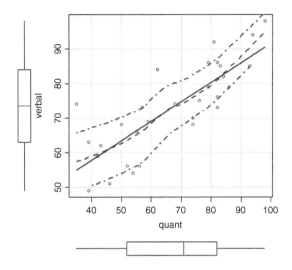

The `scatterplot()` function generates for us a **line of best fit**, which as we will see, is the **least-squares line** for predicting `verbal` based on knowledge of `quant`. A confidence envelope around the line is also generated automatically, though this need not concern us just now. The important point for now is that we see that the relationship between `verbal` and `quant` is approximately **linear** in form. A general increase in `quant` appears to be related to a general increase in `verbal`. Boxplots for both `quant` and `verbal` are also generated below the axis of each variable which depicts their respective univariate distributions as opposed to the **bivariate** distribution depicted in the scatterplot.

Informally, we can use this line to make predictions. For instance, suppose you wanted to predict a person's verbal score from a quant score of 70. We could draw a vertical line up to the regression line, then read off the predicted value for verbal:

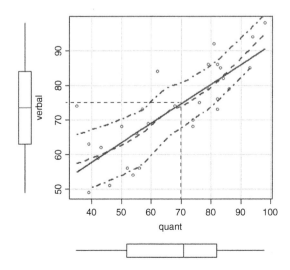

As a guesstimate then, the predicted value for `verbal` falls approximately at 75 or so for a value of `quant` equal to 70. This is a quick and fast example of how we can use regression analysis, though it doesn't begin to tell us what the method is actually doing. We now look at that in more detail beginning with a survey of ordinary least-squares.

7.2 Ordinary Least-Squares Regression

The line fit by R in the preceding plot was constructed in a very specific and unique fashion. That is, to compute it, we had to follow very precise formulas. Its exact fit is definitely not arbitrary. As mentioned, it is referred to as the least-squares regression line. But what is a least-squares line, exactly? Let's define the **population least-squares regression line** as follows:

$$y_i = \alpha + \beta x_i + \varepsilon_i,$$

where α is the population intercept of the line, and β is the population slope. Both of these are **parameters** that we will seek to estimate based on our available sample data. The first part of the equation $\alpha + \beta x_i$ is the **deterministic** part of the model, in that once we have estimated parameters for α and β, we can then make predictions on y_i. However, our predictions will not always be perfect. We need something in the equation that reveals this imperfection in prediction. That something is the **error term**, ε_i, which are the **errors in prediction**. This is the **random** or **stochastic** part of the model.

So, to recap, once we have estimated values for α and β, along with inputting a given value for x_i, we will be able to make predictions for y_i, and the extent to which our predictions are "off" or "in error," will be reflected in the error term for the model denoted by ε_i.

What is so special about the least-squares regression line? First of all, why is it called such, by the name of "least-squares?" The reason it is given this name is that the formulas used to fit the least-squares line, in terms of computing the intercept and slope, are calculated in such a way that they guarantee that the resulting line generates the best fit possible for the data in a very particular sense. The line is fit in the sense of, on average, **minimizing errors in prediction**.

We denote a predicted value for y_i by y_i'. These values of y_i' correspond to the predicted values of y_i that fall exactly on the line. However, for any prediction, we have the chance of making an error in prediction, of the form $y_i - y_i'$. Now, it stands that the sum of all these errors in prediction is equal to 0, since by definition, the least-squares line was fit in such a way that it serves as a kind of

balance or **equilibrium** for the data, a sense of a **floating average**. Hence, to prevent the sum of these deviations always equaling zero, we will **square deviations**, so that we obtain a sum of squared errors of the following form, where now sample estimates for α and β (denoted by a and b, respectively) are used in place of the corresponding population parameters:

$$\sum_{i=1}^{n} e_i^2 = \sum_{i=1}^{n} [y_i - (a + bx_i)]^2$$

It is a mathematical truth of the least-squares line that it minimizes this sum, in the sense that had we placed the line anywhere else instead of where it is, it would not have resulted in the smallest possible sum of squared errors. But what defines the line as such, precisely? How we compute a and b defines precisely the location of the line. The intercept $a_{y \cdot x}$, where the subscript $y \cdot x$ denotes "y given x as a predictor," is computed by

$$a_{y \cdot x} = \bar{y} - b_{y \cdot x} \bar{x},$$

where $\bar{y}$ is the mean of y_i, $\bar{x}$ is the mean of x_i, and $b_{y \cdot x}$ is the slope of y_i on x_i, or equivalently, the slope for y given x as a predictor. It can be proven mathematically that computing the intercept this way, and using this as an estimate of the population parameter α guarantees for us that we are computing the intercept that will serve to generate the least-squares line, that is, to keep the sum of squared errors to a minimum value.

The computation for the slope is given by

$$b_{y \cdot x} = \frac{\sum_{i=1}^{n}(x_i - \bar{x})(y_i - \bar{y})}{\sum_{i=1}^{n}(x_i - \bar{x})^2}.$$

Computing the slope in this fashion, in combination with how the intercept was computed, guarantees the line we are fitting is the least-square line, that is, the line that will minimize the sum of squared errors, and thus on average, guarantees in a sense that our predictions **overall and on average**, will be correct. Though as mentioned, for any given prediction, we stand to make a small-to-large error in prediction of the form $e_i = y_i - y_i'$, where e_i now denotes an error in prediction in the sample.

We can also obtain the earlier plot of `verbal` as a function of `quant` using **ggplot2**:

```
> ggplot(iq.data, aes(quant, verbal)) +
+ geom_point()
```

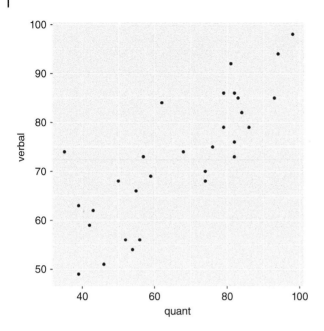

We now run the regression of `verbal` on `quant` in R:

```
> fit <- lm(verbal ~ quant)
> fit

Call:
lm(formula = verbal ~ quant)

Coefficients:
(Intercept)          quant
    35.1177         0.5651
```

The linear regression is specified by `lm()` and the function statement is `verbal ~ quant`. Calling the object `fit`, `lm()` reports selective output for only the intercept and slope estimates. We can use `coef()` to get the estimates as well:

```
> coef(fit)
(Intercept)          quant
  35.117653       0.565105
```

The more complete statement to the above would have been to first identify the data frame, and request coefficients from it. So we would have gotten the same thing by coding `fit$coef`.

From this output, we can now state the estimated regression line:

```
predicted verbal = 35.117653 + 0.565105(quant)
```

To get predicted values, we simply insert a value for quant and compute the equation. For example, for a quant score of 10, predicted verbal score is estimated as:

```
predicted verbal = 35.117653 + 0.565105(10)
                 = 35.117653 + 5.65105
                 = 40.769
```

That is, for a quant score equal to 10, our regression equation predicts a verbal score equal to 40.769. We will see soon how we can generate these fitted values in R instead of computing them manually.

Notice that the intercept of 35.117653 is the predicted value for verbal when quant is equal to 0. That is,

```
predicted verbal = 35.117653 + 0.565105(0)
                 = 35.117653
                 = 35.117653
```

That is, for a score of 0 on quant, our best prediction of verbal is equal to the intercept value of 35.117653.

As usual, we can get a much richer summary of the results via the summary() function:

```
> summary(fit)

Call:
lm(formula = verbal ~ quant)

Residuals:
     Min        1Q     Median       3Q       Max
 -11.6333   -6.5656   -0.0252    5.4103   19.1037

Coefficients:
             Estimate Std. Error t value Pr(>|t|)
(Intercept)  35.11765    5.39147   6.514 4.65e-07 ***
quant         0.56511    0.07786   7.258 6.69e-08 ***
---
Signif. codes:  0 '***' 0.001 '**' 0.01 '*' 0.05 '.' 0.1 ' ' 1
```

```
Residual standard error: 7.779 on 28 degrees of freedom
Multiple R-squared:  0.6529,    Adjusted R-squared:  0.6405
F-statistic: 52.68 on 1 and 28 DF,   p-value: 6.688e-08
```

Let's unpack the output:

- `Residuals` provides a 5-point summary of the distribution of differences between observed and fitted values. We will discuss residuals more thoroughly a bit later in the context of multiple regression.
- The `Intercept` is the expected mean of `verbal` for when `quant` = 0. That is, as mentioned, it is the predicted value for `verbal` when one has a measure of 0 on `quant`. Though a significance test on the intercept is rarely of value in this particular context, R nonetheless generates one by default. The estimate of the intercept equal to 35.11765 divided by the estimated standard error of 5.39147 yields a *t* value of 6.514, which in turn yields a *p*-value of 4.65e−07. Hence, we would say the intercept is statistically significant.
- The coefficient for `quant` is of much greater interest, and is equal to 0.56511. Its associated standard error is equal to 0.07786. Recall that the standard error of a statistic is a measure of how precisely the statistic was estimated. Smaller standard errors generate more power against the null hypothesis, and larger ones make rejecting the null more difficult. For our data, the slope for `quant` is statistically significant with $p = 6.69e−08$. The interpretation of the coefficient is as follows: **for a one-unit increase in quant, we expect, on average, verbal to increase by 0.56511 units**. Note that it would be incorrect to conclude that for a one-unit increase in `quant`, `verbal` increases by 0.56511 units, since this does not take into account the "messiness" of the data, it only considers the perfect linear form we fit to this messy surface. By adding the qualifiers "expect" and "on average," we are communicating that prediction is not perfect, and that the line represents an estimated function amid this imperfect swarm of points.
- `Residual standard error: 7.779 on 28 degrees of freedom` – this corresponds to the square root of the residual variance from the ensuing ANOVA table from the regression (we print this table to follow and will discuss its contents). That is, when we square 7.779, we obtain the residual variance of 60.5 appearing in the ANOVA table under `Mean Sq Residuals`.
- `Multiple R-squared: 0.6529` – this is the squared correlation between observed and predicted values of y_i. It is referred to as "multiple" *R*-squared simply to denote its most general possibility for when we have more than a single predictor. That is, in a multiple regression setting (which we will survey soon), it is truly and definitively "multiple" *R*-squared. In the current setting, however, with only a single predictor, it would suffice to call it simply "*R*-squared."

To demonstrate its computation, let's generate this number manually in R, where ^2 represents squaring:

```
> cor(fitted(fit), verbal)^2
[1] 0.6529401
```

- Equivalently, because this is a simple linear regression, meaning that there is only a single predictor, multiple *R*-squared of 0.6529 can be conceptualized as the squared correlation between observed values on `verbal` and observed values on `quant`. That is, it is simply the squared correlation between verbal and quant, as we can readily demonstrate:

```
> cor(verbal, quant)^2
[1] 0.6529401
```

- Adjusted *R*-squared of 0.6405 penalizes *R*-squared somewhat in the sense of potentially fitting more parameters than absolutely necessary. In this sense, it is a more conservative estimate of what *R*-squared actually is in the population. We discuss R^2_{Adj} a bit further, shortly.
- Finally, `F-statistic: 52.68 on 1 and 28 DF, p-value: 6.688e-08` – this is obtained *F* of 52.68, evaluated on 1 and 28 degrees of freedom, with associated *p*-value of 6.688e−08; hence, it is deemed statistically significant. This *F* also appears in the ANOVA table that follows.

```
> anova(fit)
Analysis of Variance Table

Response: verbal
          Df Sum Sq Mean Sq F value    Pr(>F)
quant      1 3187.3  3187.3  52.678 6.688e-08 ***
Residuals 28 1694.2    60.5
- - -
Signif. codes:  0 '***' 0.001 '**' 0.01 '*' 0.05 '.' 0.1 ' ' 1
```

But what is an **Analysis of Variance Table** doing in regression output? Though understanding this requires a bit more theory than we can get into in this book, you should realize that ANOVA, the **procedure** of analysis of variance surveyed earlier in the book, is for the most part but a special case of the wider regression model framework. That is, both the analysis of variance and regression accomplish something remarkably similar, and that is to partition variability into that accounted for by the model and that which is not. In the above ANOVA table, notice we have sources of variation (i.e. sums of squares) for quant and Residuals. In brief, the partition of these sources of variability into 3187.3 for quant and 1694.2 for Residuals is a partition of variability into that accounted for by the regression model (i.e. the regression line) and that unaccounted for by the model. Notice that if we take the ratio SS quant to the sum SS quant + SS residuals, we get:

$$\frac{\text{SS quant}}{\text{SS quant} + \text{SS residuals}} = \frac{3187.3}{3187.3 + 1694.2} = 0.6529$$

This number of 0.6529 is equal to the multiple R-Squared we computed earlier, though we described it as the squared correlation of observed and predicted values. Either way we define, it produces the same number. In understanding regression, however, it's helpful to see this number as the above ratio because it expresses so concisely one way of conceiving regression, that of partitioning variability into that accounted for by the model (i.e. SS quant) and that unaccounted for by the model (i.e. SS residual). The proportion therefore accounted for by the model must be SS quant over the **total variation possible**, which must be SS quant + SS residuals.

7.3 Adjusted R^2

We just defined R-squared as the squared correlation between observed and predicted values from our model, or, equivalently as SS regression over SS total. R-squared is a very popular measure of model fit, and is regularly reported in results sections of scientific reports. However, one drawback of R-squared is that it will **always increase as we add complexity to the model** (assuming at least some contribution from the new variable). That is, as we add more predictors to a model, even if those predictors are weak predictors, R-squared will increase. It can't go down.

For this reason and others, other measures of model fit have been proposed over the years to account for R-squared's shortcomings, and attempt to incorporate a sense of **model complexity** into their calculations. One such measure is R_{Adj}^2, pronounced "adjusted R-squared." What R_{Adj}^2 seeks to do is incorporate model complexity into its measure, and in a sense, "punish" the user for making the model more complex than it needs to be.

There are a few different ways of defining R_{Adj}^2. A fairly popular version is the following:

$$R_{Adj}^2 = 1 - \left(1 - R^2\right) \left(\frac{n-1}{n-p}\right),$$

where n is the number of observations in the model, and p is the number of parameters fit in the model, which includes the intercept term. But what does R^2_{Adj} accomplish, exactly? Since it is a fact that R^2 can only increase as more variables are added to a model, as mentioned, R^2_{Adj} attempts to mitigate this fact. That is, in the mind of R^2_{Adj}, it seems more reasonable that a measure of fit should only increase if the added variables are worthwhile adding to the model in the sense of whether they make a fair contribution to the model. This is exactly what R^2_{Adj} seeks to accomplish, to account for the possibility that added variables are contributing more **noise** than actual **variance explained**. When we conduct a simple or multiple regression then, we can expect R^2_{Adj} to be smaller than R^2.

7.4 Multiple Regression Analysis

In the regression model featured thus far, we have had only a single predictor. In our example just performed in R, `verbal` was hypothesized as a function of `quant` only, and no other predictors. Often in research, however, we would like to model two or more predictors **simultaneously**. That is, we would like to include more than a single predictor into our regression model. When we are including two or more predictors into the same regression model, the model is by definition a **multiple regression**, given by:

$$y_i = \alpha + \beta_1 x_{1i} + \beta_2 x_{2i} + \cdots + \beta_p x_{pi} + \varepsilon_i,$$

where α is again the intercept parameter of the equation, and β_1 through β_p are what are known as **partial regression slopes**. We will have as many partial regression slopes as there are predictors in our model, hence the need for subscripting with p for both β_p and x_{1i} through x_{pi}. The model still retains one error term ε_i, which as was the case for simple linear regression, is the degree to which the deterministic portion of the model (i.e. $\alpha + \beta_1 x_{1i} + \beta_2 x_{2i} + \cdots + \beta_p x_{pi}$) cannot account for full and accurate prediction of y_i.

We now demonstrate a full multiple regression analysis in R. Through unpacking its output, we note its differences with simple regression. To make it a multiple regression, instead of only `quant` predicting `verbal`, let's now add `analytic` to the model as well. That is, we will model both `quant` and `analytic` simultaneously as predictors of `verbal`:

```
> fit.mr <- lm(verbal ~ quant + analytic)
> summary(fit.mr)

Call:
lm(formula = verbal ~ quant + analytic)
```

```
Residuals:
    Min       1Q    Median       3Q       Max
-11.8924  -6.4854  -0.7326    5.2517   19.0404


Coefficients:
            Estimate Std. Error t value Pr(>|t|)
(Intercept)  32.1709     5.8862   5.465 8.74e-06 ***
quant         0.4281     0.1378   3.107  0.00442 **
analytic      0.1696     0.1414   1.200  0.24055
- - -
Signif. codes:  0 '***' 0.001 '**' 0.01 '*' 0.05 '.' 0.1 ' ' 1


Residual standard error: 7.718 on 27 degrees of freedom
Multiple R-squared:  0.6705,    Adjusted R-squared:  0.6461
F-statistic: 27.47 on 2 and 27 DF,   p-value: 3.096e-07
```

We unpack the details of the model fit. As we did for the simple linear regression, we will skip the `Residual` output for now, and proceed directly to the interpretation of the coefficients and model fit indicators:

- The partial regression coefficient for `quant` is estimated to be 0.4281, and is statistically significant at $p = 0.00442$. We interpret this coefficient in a similar fashion as we did in simple linear regression, but with an important caveat; we must also mention that we have included `analytic` in the model. Thus, our interpretation is: for a one-unit increase in `quant`, we expect, on average, `verbal` to increase by 0.4281 units, given the inclusion of `analytic` in the model, or "partialling out" or "controlling for" `analytic` in the model.

- Likewise, we interpret the coefficient for `analytic` in a similar manner as `quant`, though in this case, it is not statistically significant. When interpreting `analytic`, we would say, "for a one-unit increase in `analytic`, we expect, on average, `verbal` to increase by 0.1696 units," given the inclusion of `quant` in the model, or "partialling out" or "controlling for" `quant` in the model.

- The `Residual standard error` of 7.718 is again, as it was in simple linear regression, the square root of MS residual from the corresponding ANOVA table, and gives a sense of how much unexplained variance exists in the regression model:

```
> anova(fit.mr)
Analysis of Variance Table
```

```
Response: verbal
          Df Sum Sq Mean Sq F value     Pr(>F)
quant      1 3187.3  3187.3 53.5057 7.218e-08 ***
analytic   1   85.8    85.8  1.4401    0.2405
Residuals 27 1608.4    59.6
---
Signif. codes:  0 '***' 0.001 '**' 0.01 '*' 0.05 '.' 0.1 ' ' 1
```

- Multiple R-squared: 0.6705 is, as was true in simple linear regression, the squared bivariate correlation between observed and predicted values on y_i. As we did in simple regression, we can confirm this calculation quite easily:

```
> cor(verbal, fitted(fit.mr))^2
[1] 0.670514
```

Notice that because this is multiple linear regression where we have more than a single predictor variable, multiple R^2 cannot be computed as the bivariate correlation between verbal and a given predictor, since there are numerous predictors in the model. Rather, multiple R^2 is the correlation between verbal and an optimal linear combination of the predictors in the model. Similar to how we did in simple regression, we can also compute multiple R^2 directly from the ANOVA table as SS quant + SS analytic divided by SS total, where SS total is equal to SS quant + SS analytic + SS residual. The computation of multiple R^2 is thus $(3187.3 + 85.8) / (3187.3 + 85.8 + 1608.8) = 0.6705$.

The value for Adjusted R-squared is 0.6461, which, as was the case in simple linear regression, penalizes R-squared somewhat in the sense of potentially fitting more parameters to the model than absolutely necessary. That is, we expect R^2_{Adj} to be smaller than multiple R^2, and note that it is for our data. Whereas R^2 is equal to 0.6705, R^2_{Adj} is a bit smaller at 0.6461. Finally, the F statistic of 27.47 on 2 and 27 degrees of freedom is statistically significant with an associated p-value of 3.096e−07.

Multiple linear regression differs from simple linear regression by including more than a single predictor into the model. Multiple regression assesses the contribution of predictors while partialling out the influence of other predictors. Instead of interpreting the slope, we interpret the partial regression slope in multiple regression, and the interpretation of any coefficient must be done noting that it is being interpreted in the presence of the other variables simultaneously included into the model.

7.5 Verifying Model Assumptions

Recall that most statistical models come with them a set of assumptions, that if not at least somewhat satisfied, can cast doubt on the suitability of the model as well as its ability to infer to the population from which the sample was drawn. The assumptions that need to be verified in regression typically fall into the following categories:

1) **Normality of Errors.** Since we do not have the actual errors from our regression, we will instead inspect the **residuals** from the model, and verify that they are approximately normally distributed. We can verify the normality of errors assumption for our multiple regression using a Q–Q plot and a histogram, where resid(fit.mr) generates the residuals for the model object fit.mr:

```
> library(car)
> qqPlot(fit.mr)
> hist(resid(fit.mr))
```

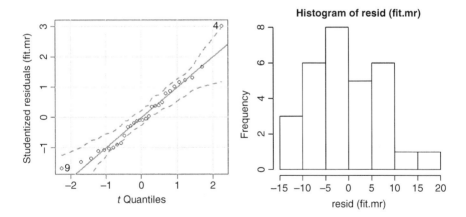

We see from the above plot that the **studentized residuals** more or less follow a normal pattern. Studentized residuals are constructed by deleting observations one at a time, and each time refitting the regression model on remaining observations (for details, see Fox, 2016). The normality is generally supported as well by the accompanying histogram on the raw residuals. A few observations deviate somewhat from a perfect normal curve, but probably not enough to be too concerned about in terms of violating normality.

2) **Homoscedasticity of Errors**. This specifies that the distribution of errors is approximately the same for each conditional distribution of the predictors. Again, since we do not have the actual errors available, we will study plots of residuals to confirm that this assumption is more or less satisfied by inspecting the pattern of residuals resulting from the regression. To verify the assumption, it suffices to plot the residuals against predicted or "fitted" values from the regression:

```
> plot(fitted(fit.mr), resid(fit.mr))
```

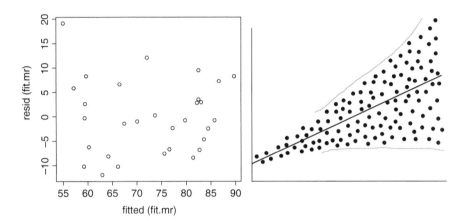

fitted (fit.mr)

If the assumption of homoscedasticity is satisfied, then residuals should more or less distribute themselves relatively evenly across the plot. That is, if you draw a horizontal line at approximately zero, there should not be any discernible pattern among the residuals, linear or otherwise, for any sub-sections of the plot (e.g. even if half of the plot exhibited a linear trend it could indicate a problem). Problems with the assumption are often indicated by a fan-shape distribution (an example appears to the right of the plot of such a violation). Instead of raw residuals, one can also plot standardized residuals via `rstandard()`. So for our data, we would obtain them by `rstandard(fit.mr)`. Plotting the aforementioned studentized residuals is often recommended as well (see Fox, 2016, for details).

3) **Absence of Outlying or Influential Observations**. This assumption specifies that there are no observations in the data that are, generally speaking, very **distant** or **influential** from the others. To verify this assumption, computing **hat values** and **Cook's distance** is helpful. These are measures of general leverage and influence, respectively. We can obtain a plot of hat values by the following:

```
> plot(hatvalues(fit.mr))
```

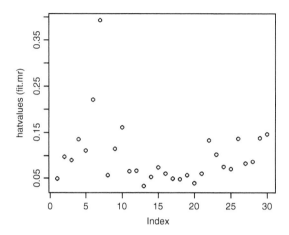

We can see from the plot that one hat value in particular stands out. To find out what observation that is, we can request the largest element of the generated hat values:

```
> which.max(hatvalues(fit.mr))
7
```

We can detect from the plot observation 7, and see that it is rather extreme. There is no strict cut-off for determining what are small versus large hat values. A useful rule-of-thumb cut-off is that any values that are greater than twice their average may be of concern. The average hat value is given by $(p + 1)/n$, where p is the number of predictors, and n is the sample size. For our data, $(p + 1)/n = (2 + 1)/30 = 0.1$, and hence any values greater than $2(0.1) = 0.2$, may be worth looking into. Only a couple of observations in our data exceed this cut-off.

We can get Cook's d in R by the following:

```
> cooks.distance(fit.mr)
           1            2            3            4
3.149727e-02 6.590444e-05 4.107587e-03 3.649510e-01
           5            6            7            8
2.654237e-02 1.399952e-01 2.295872e-01 2.361968e-02
           9           10           11           12
1.144384e-01 1.327193e-01 8.411296e-04 4.150144e-04
          13           14           15           16
2.211863e-05 1.424985e-02 6.921672e-02 2.195964e-02
          17           18           19           20
1.564389e-03 1.612958e-04 2.486268e-02 1.051286e-02
          21           22           23           24
3.122560e-03 1.242057e-02 4.158477e-03 2.212432e-02
          25           26           27           28
```

```
9.578965e-03 5.493941e-02 4.968353e-02 5.014579e-03
          29              30
7.071060e-02 5.970152e-04
```

Typically, we look for relatively large values of Cook's *d*. Scientific notation is sometimes a blessing because it condenses information neatly. However, other times, it simply looks confusing. We can turn off scientific notation using options(scipen=999) and get Cook's *d* in regular notation (we print only a few cases below to demonstrate):

```
> options(scipen=999)
> cooks.distance(fit.mr)
          1            2            3            4
0.03149726525 0.00006590444 0.00410758660 0.36495100729
```

Inspecting the values of Cook's *d*, we see that observation 4 has a relatively large value compared to the rest (i.e. 0.36). Rigorous cut-offs here are generally not recommended. For our data, no values are likely of any concern. One can also use influence.measures() to obtain hat values, Cook's *d*, and a host of other measures useful for detecting outlying observations.

We have only scratched the surface here with regard to diagnostics for regression. In most cases, simple checks as we did above will be adequate for most data to ensure there are no serious violations. However, one can delve much further into residual analysis and obtain a whole slew of additional diagnostics, including **partial residual plots** that plot the relationship between a given predictor and the response while considering all other predictors in the regression model. Fox (2016) is the authority when it comes to diagnostics in regression (and most other topics in regression as well), and should be consulted for further information. When assumptions are violated, **bootstrapping** coefficients also becomes a viable alternative where in essence the sampling distribution of the statistic is simulated based on repeated sampling from which parameter estimates are obtained. Bootstrapping is beyond the scope of this book. For details on how bootstrapping works along with applying it in R, see James et al. (2013).

Don't FORGET! *Assumptions in regression that require verification include normality, homoscedasticity, and the absence of outlying or influential observations. These can be verified using residual plots as well as computing measures such as hat values and Cook's d, among others.*

7.6   Collinearity Among Predictors and the Variance Inflation Factor

Though we expect predictors in multiple regression to be correlated to some degree, extreme correlation among them is not favorable. What is extreme? There is no set rule, however, variables exhibiting bivariate correlations upward of 0.90 or greater (as a crude guideline) can be increasingly considered **collinear**. The problems with collinearity among variables are twofold:

1) Substantively, if one variable is highly collinear with another, then they are accounting for similar proportions of variance (think of overlapping Venn diagrams), and hence though they are not replications of one another, it suggests that one of the two variables may be sufficient for study, and the other safely discarded. Now that's a bold statement, and like so many things, the issue is more complex than that. Do not simply discard one predictor based on low to moderate collinearity. On a substantive scientific level, you may still wish to include both variables, and it is the drive of the research question that should sometimes take precedence in this case, not the statistical issue entirely.

2) Variables that are collinear can pose mechanical and statistical challenges to the regression. In the event that one is an exact linear combination of the other, then linear algebra tells us that these two variables (represented by vectors) will be **linearly dependent**. Linear dependence in linear algebra implies a matrix of less than full rank (which means not all columns are linearly independent), and will halt the classic regression program. We can easily demonstrate this with a fictitious example in R, where we construct two predictors, p1 and p2, but where one will be an exact linear combination of the other. We also create a response variable y, but that isn't of most importance just now:

```
> p1 <- c(2, 4, 6, 8, 10)
> p2 <- c(4, 8, 12, 16, 20)
> y <- c(12, 15, 8, 4, 1)
```

Notice that p2 is simply each element of p1 multiplied by 2. This is only one way in which a linear combination can arise. We say that p2 is a linear combination of p1. In this case, it is a clear scalar multiple as well. Let's try running the regression with these two predictors in the model, presenting only the most essential output:

```
> model <- lm(y ~ p1 + p2)
> summary(model)
```

```
Coefficients: (1 not defined because of singularities)
            Estimate Std. Error t value Pr(>|t|)
(Intercept)  17.9000     2.7815   6.435  0.00761 **
p1           -1.6500     0.4193  -3.935  0.02924 *
p2                NA         NA      NA       NA
---
Signif. codes:  0 '***' 0.001 '**' 0.01 '*' 0.05 '.' 0.1 ' ' 1

Residual standard error: 2.652 on 3 degrees of freedom
Multiple R-squared:  0.8377,    Adjusted R-squared:  0.7836
F-statistic: 15.48 on 1 and 3 DF,  p-value: 0.02924
```

The key message in the above is 1 is not defined because of singularities appearing next to Coefficients. Notice that the coefficient for p2 is NA (not available). This is due to the extreme collinearity introduced into the model by having p2 be a scalar multiple of p1. What R did in this case was to drop p2 entirely from the model, as we can confirm when we run the model without p2:

```
> model.1 <- lm(y ~ p1)
> summary(model.1)
```

```
Coefficients:
            Estimate Std. Error t value Pr(>|t|)
(Intercept)  17.9000     2.7815   6.435  0.00761 **
p1           -1.6500     0.4193  -3.935  0.02924 *
—
Signif. codes:  0 '***' 0.001 '**' 0.01 '*' 0.05 '.' 0.1 ' ' 1

Residual standard error: 2.652 on 3 degrees of freedom
Multiple R-squared:  0.8377,    Adjusted R-squared:  0.7836
F-statistic: 15.48 on 1 and 3 DF,  p-value: 0.02924
```

Notice that the above coefficients mirror those generated for p1 in the collinear model. To help identify problems of collinearity with predictors, we can compute something known as the **variance inflation factor**, or **VIF** for short, which is computed for each predictor entered into the model. Understanding how VIF works is beyond the scope of this book (see Denis, 2016, for details), so we cut right to the chase; relatively large values for VIF for a predictor are indicative that that predictor might be collinear with other predictors in the model. They are a measure of how much the variance of a predictor (in terms of its variance or standard error) is "inflated" due to being correlated with other predictors in the model. The minimum VIF can achieve is 1 and signifies no inflation. As VIF gets larger, it is indicative of a potentially more serious problem. We can try computing VIF for our predictors in the original model:

```
> library(car)
> vif(model)
Error in vif.default(model) : there are aliased
coefficients in the model
```

In this particular case, since p2 is an exact scalar multiple of p1, R reports that the model contains aliased coefficients and does not report a VIF value at all. Let's ease the degree of collinearity by modifying p2 slightly. We will change the final value in p2 from 20 to 21, and rerun the model:

```
> p2 <- c(4, 8, 12, 16, 21)

> model <- lm(y ~ p1 + p2)
> summary(model)

  Coefficients:
              Estimate Std. Error t value Pr(>|t|)
(Intercept)    17.500      3.940   4.441   0.0471 *
p1              0.450     10.694   0.042   0.9703
p2             -1.000      5.087  -0.197   0.8623
```

We note that the model was able to run, and coefficients for p1 and p2 were estimated. Now, let's compute VIF values for the predictors:

```
> vif(model)
 p1  p2
442 442
```

The VIF values are extremely large due to the collinearity. Of course, in practice, VIF values will seldom be this high, but values larger than 5–10 should cause you to pause and consider whether it might be worth dropping one or more of the predictors in the model. Substantively, from a scientific point of view, VIFs certainly of 5 or higher should cause you to ponder your data a bit further to assure yourself you're not measuring similar constructs with two predictors. It may be possible you have two items in your questionnaire that are essential duplicates, for example one of which asks respondents how happy they are, and another how much they are experiencing happiness. Large VIFs not only help us spot **statistical** issues of dependency, but they can also help us spot **scientific** issues.

7.7   Model-Building and Selection Algorithms

In this chapter thus far, we have surveyed the methods of simple and multiple linear regression. When we are conducting a simple linear regression, there is, of course, only one possibility for the model, since there is only a single predictor. However, when we are conducting multiple linear regression, thereby having more than a single predictor, the question then becomes that of choosing or selecting the **best** model. The problem is how to best define best! Which model shall we adopt? Should we adopt a model with very few predictors, or should we adopt a model with many more predictors? For instance, if we have 20 predictors available, should we use all of them, or only a subset of them in our quest to define the best model? How should that **subset** be chosen as to optimize some function of the data?

Surprising to most novices in the area, these questions are not easy for even the most experienced of researchers or data analysts. As noted by Izenman (2008, p. 143), "Selecting variables in regression models is a complicated problem, and there are many conflicting views on which type of variable selection procedure is best." Indeed, as we will see, how a "best" model is arrived at will often depend on **substantive** and **scientific** issues as it will on statistical criteria. We survey both in this section. Generally, a guiding principle in model selection is that **simplicity rules the day**. That is, given a choice between a more complex model and a simpler one, all else equal, the simpler one is usually preferable to the more complex. This principle often goes by the name of **Occam's razor** and is applicable not only to statistical models, but to virtually all narrative explanations. For example, if one could explain depressive symptoms based on fluctuating blood sugar levels equally well with that of a complex parental dynamic, the blood sugar theory would likely be preferable to the more complex theory. Of course, that doesn't imply the blood sugar theory is "correct" (recall that no theory is generally ever "correct" or "true"), only that in terms of explanation, the simpler theory is usually preferable if it accounts for similar amounts of variance as the more complex one. The same applies to statistical models, in that we typically prefer models that are simple if they account for similar amounts of variance as do more complex ones.

7.7.1   Simultaneous Inference

The first and most obvious way to select a model is to simply include all available predictors into the model and assess model fit based on this complete model. We may call this approach **full entry** or **simultaneous** regression. In this approach, we build the regression model by simultaneously estimating all parameters at the same time. In our previous example of multiple regression of `verbal` on `quant` and `analytic`, it would mean simultaneously modeling `quant` and `analytic` as predictors of `verbal`. That is, simultaneous inference in this case would mean running the model we already ran where both predictors were entered into the model at the same time.

When should simultaneous regression be used? When the researcher has a relatively strong theory-driven regression to begin with, simultaneous regression is often the preferred choice. It is the regression model we first think of when we consider a multiple regression, and for most research scenarios, it is the ideal way to proceed. We could even say that it is not really a "model selection" procedure at all as the phrase is most offered referred to, since there is only a **single model** to test regardless of how many predictors we are entering into the model. In one's exploration of running a suitable model for their variables, researchers should usually begin by considering simultaneous regression as the default option unless there is reason to perform a different competing model-building algorithm. After all, most researchers are presumably coming to the research area with some prior knowledge of the kind of model they would like to obtain empirical support for, and the building of a regression model should, in a strong sense, be a demonstration of that prior knowledge.

However, there are instances where full-entry or simultaneous regression may not be considered the best option for model-building, and where the researcher may be more interested in adopting a more complex algorithm to building his or her regression model. For example, at times a researcher may wish to prespecify the exact order of entry of variables into the model instead of adding them in simultaneously. For that, a hierarchical regression may be called for. We survey that possibility next.

7.7.2 Hierarchical Regression

In **hierarchical regression**, in contrast to simultaneous regression where all predictors are entered into the model at the same time, a researcher usually has a designated **prespecified order** in which he or she would like to enter predictors into the model. This order of entry is usually **theoretically driven**, based on the prior knowledge base of the researcher. For example, given a response variable y_i and a set of predictors x_1 through x_p, a researcher may wish to enter one or more predictors first, as a set, as to partial out their variance initially, before examining the contribution of other predictors. For instance, referring to the IQ data once more, perhaps a researcher knows in advance from prior research that in predicting verbal, both quant and analytic likely have a certain degree of "overlap." Hence, in wanting to know the predictive power of quant, the researcher feels it necessary to first "control for" analytic. So, in **Block 1**, the researcher would enter analytic, then in **Block 2**, enter quant to observe the **increment in variance** explained. This is an example of hierarchical regression because the order of entry of predictors was predetermined by the researcher. That is, it was based on theory, rather than purely statistical criteria alone.

As another example, suppose a researcher would like to investigate the predictive power of cognitive-behavior therapy on depression (i.e. depression is the response, cognitive behavioral therapy [CBT] is the predictor), but all the while knowing in advance that depression differs by gender. The researcher would presumably have knowledge of this based on prior work on associating depression with gender. In such a case, the researcher may choose to include gender first into the model at Step 1, then at Step 2 enter CBT to observe its impact over and above that of gender. Again, in this sense, the researcher hopes to "control for" gender while then being able to observe the effect of CBT on depression.

It is imperative to note that in hierarchical regression, the idea of controlling for a predictor by first entering it into the model is not at all the same as true **experimental control**. That is, when gender was included in the first step of the model in our prior example, we were not actually truly controlling for it in the sense of real control. Rather, the control in question was merely a **statistical control** in the sense of partialling out variability due to gender before considering the impact of CBT over and above gender. It cannot be emphasized enough that you must not associate statistical control with that of experimental control, as the two are not one and the same.

Statistical control is not the same as true experimental control. When we say we are controlling for a variable in a hierarchical regression by entering it at a particular step (often the first step of the model-building procedure), all we are doing is partialling out that variable's variance, and not actually controlling for it in the experimental sense of the word. Hence, never associate statistical control with that of experimental control, as they are not one and the same.

7.7.2.1 Example of Hierarchical Regression

We demonstrate a hierarchical regression using the GAF data, which stands for **Global Assessment of Function**, taken from Petrocelli (2003). GAF is an overall score that reveals a sense of how an individual is functioning overall in their day-to-day life, where higher scores on GAF (gaf) indicate better functioning than do lower scores. Predictors for the model include scores on **age** (age), **pretherapy**

depression score (pretherapy, higher scores are better), and **number of therapy sessions** (n_therapy). Our basic functional model is thus the following:

$$GAF = age + pretherapy\ depression + number\ of\ therapy\ sessions$$

We read our data into R:

```
> gaf.data <- read.table("gaf.txt", header = T)
> gaf.data
   gaf age pretherapy n_therapy
1   25  21         52         6
2   25  19         61        17
3    0  18         50         5
4   43  30         57        31
5   18  27         54        10
6   35  39         55        27
7   51  36         61        11
8   20  20         52         7
9   16  23         51         8
10  47  35         55        10
```

We first evaluate a model in which we are interested in first controlling for age. That is, we first enter age into the model at step 1:

```
> attach(gaf.data)
> gaf.fit <- lm(gaf ~ age)
> summary(gaf.fit)

Call:
lm(formula = gaf ~ age)

Residuals:
    Min      1Q  Median      3Q     Max
-13.657  -9.196   4.359   7.617   9.784

Coefficients:
            Estimate Std. Error t value Pr(>|t|)
(Intercept) -15.6806    12.1427  -1.291  0.23264
age           1.6299     0.4369   3.731  0.00578 **
---
Signif. codes:  0 '***' 0.001 '**' 0.01 '*' 0.05 '.' 0.1 ' ' 1
```

```
Residual standard error: 10.19 on 8 degrees of freedom
Multiple R-squared:  0.635,     Adjusted R-squared:  0.5894
F-statistic: 13.92 on 1 and 8 DF,  p-value: 0.00578
```

We see that age is statistically significant, $p = 0.00578$, with multiple R-squared for the model at 0.635. Recall, however, that our purpose was not to evaluate age. Our purpose was simply to enter it into the model first, so that we may "control for it" and then evaluate the impact of future predictors after controlling for age. We now enter age and pretherapy in step 2:

```
> gaf.fit <- lm(gaf ~ age + pretherapy)
> summary(gaf.fit)

Call:
lm(formula = gaf ~ age + pretherapy)

Residuals:
    Min      1Q  Median      3Q     Max
-8.8399 -7.2939 -0.0415  6.6845  9.2966

Coefficients:
             Estimate Std. Error t value Pr(>|t|)
(Intercept) -102.7837    39.6255  -2.594   0.0357 *
age            1.2671     0.3888   3.259   0.0139 *
pretherapy     1.7669     0.7786   2.269   0.0575 .
- - -
Signif. codes:  0 '***' 0.001 '**' 0.01 '*' 0.05 '.' 0.1 ' ' 1

Residual standard error: 8.265 on 7 degrees of freedom
Multiple R-squared:  0.7897,     Adjusted R-squared:  0.7297
F-statistic: 13.15 on 2 and 7 DF,  p-value: 0.004263
```

A couple of things worth noting from our updated model now including pretherapy:

- Pretherapy is marginally statistically significant, $p = 0.0575$ after holding age constant (notice the p-value for age also changed from what it was previously, that's a different matter that we'll address later under the topic of **mediation**)
- We note that multiple R-squared is now 0.7897, up from 0.635 in the model with only age. This difference in R-squared values, of $0.7897 - 0.635 = 0.1547$, represents the **incremental variance explained** by the model.

- The key conclusion here goes like this: "After controlling for `age` in the model, we find that `pretherapy` is marginally statistically significant. That is, `pre-therapy` explains variance in `gaf` over and above `age` alone."
- Again, to reiterate, what makes this a hierarchical regression is the fact that it is the researcher who has specified order of entry of the predictors. This detail is important because soon we will discuss techniques such as forward and stepwise regression, in which it is the **algorithm** and computer that determines order of entry via statistical criteria alone, rather than by the prespecified criteria of the researcher. As we will discuss, these techniques are categorically different from the hierarchical regression just surveyed, and should not be confused with it.

7.8 Statistical Mediation

The technique of hierarchical regression is popular among social scientists in test-ing mediational hypotheses. **Mediation**, which employs hierarchical regression, was proposed by Baron and Kenny (1986), and is depicted in Figure 7.1.

In a mediation model, the IV (or "predictor") is hypothesized to predict the "dependent variable" (DV), as notated by path **c**. However, this hypothesized rela-tionship is said to be potentially "mediated" by the hypothesized mediator (MEDIATOR), which for the mediation model, implies that once MEDIATOR is added to the model, the path from IV to DV will drop from **c** to **c′**. If such path drops to 0, then **full mediation** is said to have occurred. If it is greatly reduced, but not equal to 0, then **partial mediation** is said to have occurred.

As an example of mediation, we will replicate the model we fit on the GAF data, but with a hypothesis that the predictive power of AGE on GAF is mediated by PRETHERAPY:

```
> mod.1 <- lm(gaf ~ age)
> summary(mod.1)
```

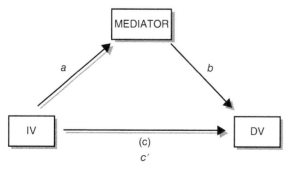

Figure 7.1 Classic single-variable mediation model.

```
Call:
lm(formula = gaf ~ age)
```

```
Residuals:
    Min      1Q  Median      3Q     Max
-13.657  -9.196   4.359   7.617   9.784
```

```
Coefficients:
             Estimate Std. Error t value Pr(>|t|)
(Intercept) -15.6806    12.1427   -1.291  0.23264
age           1.6299     0.4369    3.731  0.00578 **
---
Signif. codes:  0 '***' 0.001 '**' 0.01 '*' 0.05 '.' 0.1 ' ' 1
```

```
Residual standard error: 10.19 on 8 degrees of freedom
Multiple R-squared:  0.635,    Adjusted R-squared:  0.5894
F-statistic: 13.92 on 1 and 8 DF,  p-value: 0.00578
```

We see that age is statistically significant ($p = 0.00578$). We now hypothesize pretherapy as a mediator of this relationship. This implies that once we include pretherapy into the model, the relationship between gaf and age should greatly reduce:

```
> mod.2 <- lm(gaf ~ age + pretherapy)
> summary(mod.2)
```

```
Call:
lm(formula = gaf ~ age + pretherapy)
```

```
Residuals:
    Min      1Q  Median      3Q     Max
-8.8399 -7.2939 -0.0415  6.6845  9.2966
```

```
Coefficients:
              Estimate Std. Error t value Pr(>|t|)
(Intercept) -102.7837    39.6255   -2.594   0.0357 *
age            1.2671     0.3888    3.259   0.0139 *
pretherapy     1.7669     0.7786    2.269   0.0575 .
---
Signif. codes:  0 '***' 0.001 '**' 0.01 '*' 0.05 '.' 0.1 ' ' 1
```

```
Residual standard error: 8.265 on 7 degrees of freedom
Multiple R-squared:  0.7897,    Adjusted R-squared:  0.7297
F-statistic: 13.15 on 2 and 7 DF,  p-value: 0.004263
```

We note that once pretherapy is included in the model, the p-value for age increased to 0.0139, and pretherapy is marginally statistically significant ($p = 0.0575$). While any evidence here for mediation would at most be very weak (especially given that we likely have insufficient power for our example, the p-value for age may be simply reflecting that), some researchers would suggest the drop in p-values provides at least some very weak evidence that pretherapy mediates the relationship between GAF and age. Had the p-value for age shot up to a very large value upon entering pretherapy, then in line with mediation theory, this would have provided even more evidence of **statistical mediation**.

It should be noted that this is a relatively simple demonstration of how mediation works. Evidence of relationships among variables should also exist, according to Baron and Kenny (1986), before tests of mediation are carried out (see Howell, 2002, p. 575). Further, in estimating the mediated effect, one can conduct a variety of inferential tests that we do not have space for here (for details, see MacKinnon, 2008). Mediation analyses in social research have come a long way since Baron and Kenny, and there is an entire literature devoted to testing the mediational effect. Our example is a very simplified version.

It needs to be emphasized as well that the above is, by itself, a demonstration of how mediation can be conceived **statistically**. Whether the mediator is actually "mediating" the relationship between DV and predictor is not supported alone by simply performing a hierarchical regression as we have done. Indeed, **we fully expect p-values to change as we include more predictors; this is simply the nature of regression**. Whether there is evidence of a **physical process** in which the statistical mediation test is uncovering is an argument one must make separate from the statistical regression. The issue is complex because it highlights the fact that statistical models are not necessarily (one might even say, rarely) representations of reality. Still, many researchers are content to conclude mediation simply by witnessing a drop in path **c** to **c′** and fail to recognize that statistical criteria alone are insufficient for drawing a **substantive** conclusion of mediation. For a discussion of this issue, see Fiedler et al. (2011).

Having considered both the simultaneous and hierarchical approaches thus far, it is worth noting that they are equivalent for the final model. That is, whether a researcher fits his or her model "at once" (full-entry) or in "increments" (hierarchical), the final model that results (i.e. once all predictors have been entered when performing it hierarchically) will be the same in each model. The only difference between the two approaches is how model-building occurs "along the way" to the final model, and to reiterate, the advantage of performing the hierarchical approach is that the researcher gets to observe directly how much more variance is explained by newly entered predictors. In the simultaneous approach, one could still obtain this information (e.g. through examining squared semi-partial correlations for added terms, see Denis (2016) for details), but it is much more efficient to run a hierarchical approach in this case.

7.9 Best Subset and Forward Regression

Best subset selection involves fitting separate regressions for the pool of predictors. That is, in combining the predictors in all possible ways, best subset selection seeks to attempt to find the best model. However, as you might imagine, the number of possible models quickly becomes quite large. As noted by James et al. (2013), the number of models possible for all selections of p predictors is equal to 2^p. Hence, for the case of 20 predictors, the number of possible models is thus

$$2^p = 2^{20} = 1,048,576$$

For $p = 30$, the number possible is $2^p = 2^{30} = 1,073,741,824$. Fitting this many models and attempting to determine the "best" of this set is surely a difficult to impossible task. It would thus be advantageous if we could devise and rely on a more efficient **algorithm** that will, in some sense, **optimize** our model search without having to choose among such a large number of potential and possible models.

Forward regression, or "forward entry," is the first approach among a family of approaches where instead of the researcher making decisions about model entry directly as in hierarchical regression, a **statistical algorithm** is in command on which variables enter the regression model in some sort of succession. That is, forward regression proceeds by entering predictors based on statistical criteria alone, and no other information. The order in which the researcher would like to enter predictors, or believes they should be entered, is not of concern to forward regression algorithms, as they simply seek to **maximize statistical criteria** without regard to the prior substantive knowledge of the researcher. However, forward regression isn't as simple as that. As noted by James et al. (2013, p. 207), the number of models possible in a forward "stepwise" regression is equal to

$$1 + \frac{p(p + 1)}{2},$$

where p is again the number of predictors. For example, with 10 predictors, the number of possible models is equal to 56 using a forward regression approach. The number of possible models using a best subset approach is 1024.

7.9.1 How Forward Regression Works

We detail now how the forward regression algorithm accomplishes its task. It should be recognized that this approach might differ depending on the exact R package or software program used to do the selection. However, by understanding what we are about to survey, you will have a good understanding of how these algorithms work in general when it comes to model selection.

The first thing to note about forward regression is that once a predictor is entered into the model, it cannot be removed. Hence, if an algorithm searches for the "best" predictor among 10 predictors (we will define what we mean by "best" shortly), once that variable is selected into the model, at no time after can it be "deselected" and placed back into the original variable pool. Forward regression proceeds in a number of steps, which we outline now:

In **Step 1** of the procedure, the algorithm surveys among the host of predictors the researcher deemed theoretically available for the regression. Of these predictors, forward regression selects the predictor having the **largest bivariate correlation with the response** while also being statistically significant. If no variables meet a significance threshold such as $\alpha = 0.05$, then regardless of the size of the bivariate correlation for the given predictor and the response, the variable will not be entered into the model. How can this occur? Recall that if sample size is fairly small and thus statistical power limited, it is entirely possible that a strong predictor not be statistically significant, and hence not make its way into the model. However, if power is sufficient, and thus sample size large enough, then that predictor with the largest correlation with the response will enter into the model.

In **Step 2** of the algorithm, the computer will search among remaining predictors (i.e. the remaining $p - 1$ predictors in the set after inclusion of the first predictor) and select that predictor with the **highest squared semipartial correlation** with the response while also meeting entry criteria in terms of being statistically significant. This is what a semipartial correlation measures – it partials out the predictor already in the model (for details, see Hays, 1994). However, as already noted, the first predictor in the model remains in the model. As mentioned, once a predictor is entered, it is not removed. Hence, the "new model" now contains two predictors, the first predictor being selected that has the highest bivariate correlation with the response (and statistically significant), the second predictor having the highest squared semipartial correlation with the response (and also statistically significant). It is entirely possible that the first predictor entered is no longer statistically significant after inclusion of the second predictor. However, **forward regression will not remove the first predictor**. It remains in the model. As we will see, this approach is contrary to stepwise regression, in which predictors may be added and removed as new predictor entries become available.

Step 3 of the algorithm then, mimics Step 2, in that it searches among the remaining predictors that predictor which has the highest squared semipartial correlation with the response, in this case, partialling out predictors 1 and 2 already entered in the model. Key to all this, again, is that predictors 1 and 2 will remain in the model regardless of how much variance they explain (or their degree of statistical significance) with the inclusion of predictor 3. Forward regression, in this sense, is very "loyal" to predictors, in that once a predictor is included in the model, that predictor remains in the model regardless of "newcomer" predictors included thereafter.

Ensuing steps past the third step proceed in an analogous manner, in that the algorithm selects predictors having the largest squared semipartial correlation with the response, the partialling out now consisting of partialling all predictors already in the model.

Denis (2016) summarizes the forward regression algorithm:

> Forward regression, at each step of the selection procedure from Step 1 through subsequent steps, chooses the predictor variable with the greatest squared semipartial correlation with the response variable for entry into the regression equation. The given predictor will be entered if it also satisfies entrance criteria (significance level) specified in advance by the researcher. Once a variable is included in the model, it cannot be removed regardless of whether its "contribution" to the model decreases given the inclusion of new predictors. (p. 408)

A related approach to forward regression is that of **backward elimination**, where instead of adding predictors to the model, backward regression starts with all predictors entered, and peels away predictors no longer contributing to the model. Thus, this approach is similar in spirit to forward regression, but may result in a different final model.

7.10 Stepwise Selection

Stepwise regression can be seen as a kind of **hybrid** between forward and backward approaches. In stepwise regression, the algorithm, as it does in forward regression, searches for predictors with the highest semipartial correlation with the response. However, once a predictor is entered into the model, contrary to forward regression in which that predictor cannot be removed regardless of incoming selections, in stepwise, the existing predictor in the model is not guaranteed to remain. That is, once a second predictor is entered, the first predictor is **re-evaluated** for whether it still contributes in light of the second predictor already in the model. If it does not (i.e. if it is no longer statistically significant), then the algorithm removes it. Hence, unlike forward regression, the stepwise algorithm is not "loyal" to predictors, in that existing predictors are subjected to removal if they no longer contribute to the model given the inclusion of new predictors. Hence at each step of variable selection, variables already in the model stand the chance at being eliminated from it.

We can demonstrate a stepwise regression in R using the `stepAIC()` function. For demonstration, we test a very simple model using the IQ data, having variables `verbal`, `quantitative`, and `analytic`:

```
> library(MASS)
> iq.step.fit <- lm(verbal ~ quant + analytic, data = iq.data)
> step <- stepAIC(iq.step.fit, direction = "both")
```

The specification of `direction = both` above tells R to conduct stepwise regression, which includes a combination of forward and backward approaches. To perform a purely forward selection, we would specify `direction = forward`, and to perform a purely backward selection, we would specify `direction = backward`.

```
Start:  AIC=125.45
verbal ~ quant + analytic

           Df Sum of Sq  RSS AIC
- analytic  1        86 1694 125
<none>                   1608 126
- quant     1       575 2183 133


Step:  AIC=125.01
verbal ~ quant

           Df Sum of Sq  RSS AIC
<none>                   1694 125
+ analytic  1        86 1608 126
- quant     1      3187 4881 155
> step$anova
Stepwise Model Path
Analysis of Deviance Table
```

```
Initial Model:
verbal ~ quant + analytic

Final Model:
verbal ~ quant
```

	Step	Df	Deviance	Resid. Df	Resid. Dev	AIC
1				27	1608.4	125.45
2	- analytic	1	85.787	28	1694.2	125.01

We see that the first model tested was that of verbal ~ quant + analytic, yielding an AIC of 125.45. R then dropped the analytic variable, yielding a model having only quant as a predictor, verbal ~ quant, which gave an AIC value of 125.01, practically the same as for the full model. The Final Model is thus verbal ~ quant.

7.11 The Controversy Surrounding Selection Methods

At first glance, especially to the newcomer in statistics, selection methods seem like a great idea and the best way of proceeding with virtually any multiple regression problem. They at first can appear as the "panacea" to the modeling problem. After all, at first glance it would appear a great idea to just leave model selection to the computer so that it "figures out," in all of its complex computing capacities, the most "correct" model. Such an algorithmic approach must be the best solution, right? However, the situation is not as clear-cut as this, unfortunately, and there are many issues and problems, some statistical, others substantive, that surround selection approaches. Automation does have its downside, and not everything can be left to a computer to decide.

Statistically, selection methods based on algorithmic approaches such as forward and stepwise regression have been shown to **bias parameter estimates**, and essentially, make the resulting inferential model suspect. The degree of bias introduced varies, but as noted by Izenman (2008), can be in the order of 1–2 standard errors. Recall what this means. Standard errors are in part used to determine the statistical significance of predictors. If parameter estimates are biased, and standard errors are smaller or greater than they should be, the risk is that inferences on the resulting model may mean, well, very little, or at minimum, be inadequate and untrustworthy. As noted by Izenman (2008), obtained F statistics in stepwise approaches are likely to be related (correlated), and a maximum or minimum F on a given "step" of the procedure is likely not even distributed as an F statistic in the first place! Thus, choosing new variables based on their statistical significance and the degree to which they maximize F may be entirely misleading, and if stepwise approaches are to be used, many writers on the topic suggest treating obtained F statistics at each step with a degree of flexibility. That is, to not take the obtained F statistics too seriously, but to rather use them as a guide in selection rather than as a true and valid inferential test of the predictor.

Further, perhaps the greatest criticism of stepwise approaches is that they do not even necessarily result in the "best" model in terms of maximizing some criteria (e.g. such as R-squared), and there may be equally as good subsets among the available predictors that do just as good or even better at maximizing statistical criteria (James et al., 2013).

The above are only a couple of the cautions regarding statistical criteria when selecting the best model. Aside from statistical, there are also, and perhaps more importantly, **substantive cautions** that must be exercised when considering model selection. For one, the chosen model at the last step may not be one that has maximum **utility**. As pioneers in statistics have long lectured us going back to Bernoulli, **maximizing statistical expectation is not the same as maximizing utility**. What does this mean? An example will help. Suppose you have a set of $p = 10$ predictors available for selection, and a stepwise algorithm selects 5 of those predictors. Are those the "best" predictors for your model? Possibly, in the sense of attempting to maximize some criteria at each step of the algorithm. But that may be all. **When we bring the abstractness of statistical models down to the realities of real empirical research, we at once realize that what may be useful in the hypothetical abstract realm may not be pragmatic at the research or application level**.

For instance, suppose a university is considering predictors of graduate school success in choosing a model to best select incoming graduate students. Suppose that you are seeking a two-predictor model, and have three predictors available, namely undergraduate GPA, number of research publications, and quality of letters of recommendation. Suppose that these are subjected to a stepwise algorithm, and it turns out that GPA and quality of letters of recommendation maximize statistical criteria, but that number of research publications came in as a close third candidate for inclusion. What then is the "best" model? Based on stepwise statistical criteria alone, we would argue that the first model is best, consisting of GPA and letters. However, simply because that model may be "best" in a statistical sense does not mean it is necessarily best in a pragmatic sense, or in the sense of being the model the university will eventually adopt. Perhaps it is very time-consuming to collect and wait for letters of recommendation, and perhaps on a deeper level there is a certain degree of mistrust regarding letters, even if they are written with relative "praise" for the candidate. After all, the style in which letter writers write varies, and some letter writers always write "good" letters. Hence, even if the statistical algorithm selects recommendation letters as the second best predictor next to GPA, the university may still choose to use number of research publications instead simply because it is easier to obtain and process, and there is less uncertainty associated with them.

That is, on a statistical basis, the first model may be preferable, but on a pragmatic basis, the second model may still **dominate** in terms of utility. Similarly, in other settings, if a predictor is very expensive to recruit, regardless of whether the statistical algorithm selects it, it may not be pragmatically feasible for the designer of the model to use it. When statistical modeling meets the harsh realities of science and research, **utility matters, and model selection becomes a lot more complex than simply choosing a model that maximizes some statistical criteria**.

Hence, for these reasons and others, caution should be exercised when implementing a variable-selection algorithm, not only from a statistical perspective but also from a scientific or pragmatic one. Does that suggest variable-selection methods should not be used at all? Absolutely not. They may prove useful as a guide to choosing the correct model, so long as the user is aware of the caveats we have discussed.

This chapter only begins to scratch the surface of what is possible in regression models. Additional topics include polynomial regression, interactions and moderators, penalized regression models, and others. Excellent books on regression include Fox (2016) as well as Hastie et al. (2009), which is a very thorough (if not quite technical) text on statistical learning in general, of which regression models can be considered a special case.

Exercises

1 Distinguish between simple and multiple linear regression, and briefly discuss why performing one procedure over the other should consider both statistical and substantive elements.

2 Briefly discuss what fitting a least-squares regression line accomplishes.

3 In a simple linear regression, under what circumstance would R-squared equal 0? Under what circumstance would it equal 1.0? Explain.

4 Referring to the formula for adjusted R-squared, justify that as sample size increases, adjusted R-squared and R-squared will eventually be equal.

5 What does it mean to "bootstrap" a coefficient in regression analysis? Explain.

6 How is the interpretation of a regression coefficient in multiple regression different from that in simple linear regression? Explain why this difference in interpretation is very important.

7 Using the iris data, conduct a multiple regression in which `Sepal.Length` is a function of `Sepal.Width` and `Petal.Length`. Briefly summarize the results from the multiple regression. Then, verify the assumptions of normality and homoscedasticity of errors.

8 For the regression performed in #7, confirm using VIFs that collinearity is or is not a problem.

9 In the chapter we computed VIF values for a linear regression in which both predictors p1 and p2 were highly collinear. Manipulate the values in p1, p2, or both, in order to ease the degree of collinearity, and demonstrate that it has been eased by recomputing VIF values.

10 Interactions can be fit in regression as they can in ANOVA models, though are not covered in this book. What would an interaction mean in regression? Can you conceptualize one? (Note: this will be a very challenging question if you have never seen an interaction in regression, so brainstorm to see what you can come up with.)

8

Logistic Regression and the Generalized Linear Model

LEARNING OBJECTIVES

- Understand why binary dependent variables require a different statistical model than do continuous ones.
- Understand how logistic regression differs from ordinary least-squares regression.
- Distinguish between the odds and logged odds (logit).
- Understand what it means to exponentiate the logit to obtain the odds.
- Interpret an Analysis of Deviance Table in R for logistic regression.
- How to obtain predicted probabilities in logistic regression.
- How to run multiple logistic regression and interpret adjusted odds ratios.
- Understand the nature of the generalized linear model, and how many models can be subsumed under it.

8.1 The "Why" Behind Logistic Regression

The statistical models considered thus far in this book, for the most part, have featured dependent or response variables that have been **continuous** in nature. Recall what it means for a variable to be continuous – it implies that any value of the variable is theoretically possible and that there are no discontinuities or "breaks" between any two values. As mentioned at the outset of this book, mathematicians have a very precise way of thinking about continuity in terms of epsilon-delta definitions. More pragmatically, **a line is deemed continuous if one doesn't lift one's pencil at any point while drawing the line on a piece of paper**.

Univariate, Bivariate, and Multivariate Statistics Using R: Quantitative Tools for Data Analysis and Data Science, First Edition. Daniel J. Denis.
© 2020 John Wiley & Sons, Inc. Published 2020 by John Wiley & Sons, Inc.

For example, if IQ is to be a continuous variable, then, theoretically at least, it should be possible for anyone in our data set to obtain any score on IQ and not simply discrete pieces of it. For a continuous variable such as this, any IQ score is theoretically possible between an IQ of 100 and 101, for instance. In theory, someone could obtain a score of 100.04829, for example. That does not necessarily mean that someone in our data set did obtain such a value; it simply means that such a value is **theoretically possible** on the scale on which the variable is being measured. True continuity, as a mathematician or philosopher would define it, cannot actually exist with real empirical variables, since at some point, one needs to "cut" the value of the variable so that it is recordable into one's empirical data file. In theory, any value of the variable may be possible, but pragmatically, in terms of real measurements, all values of variables are **truncated** to some degree so that they are manageable in terms of data analysis. In fact, in many cases, so long as the variable in question has **sufficient distribution** and numerous possible values for it, it is usually enough for researchers to assume it might have arisen from a continuous distribution. Again, however, the actual numbers in one's data set for the variable could not possibly be a continuous scale itself, since between any two values are necessarily "breaks" between values.

In many cases, response variables may not be **naturally continuous**, but rather may be instead **discrete** or **categorical** in nature. For instance, survival (dead vs. alive) is obviously not a continuous variable, having only two categories. Citizenship likewise is not continuous. That is, there are no values between the categories of "American" and "Canadian" on the variable of citizenship like there are between values of 100 and 101 on IQ. Several other naturally discrete or categorical variables abound in research settings. Sometimes, an investigator may choose to deliberately transform a continuous variable into a discrete one. As an example, perhaps instead of wanting to keep IQ continuous, the researcher wishes instead to classify IQs into three categories – low, medium, and high. The IQ variable is now noncontinuous and discrete, since on this new scale, an individual can only be in one of the three categories, and no values are possible in between. That is, one cannot be between "low" and "medium" IQ for this variable.

Other times in research, a variable that at first was hoped to be continuous by the investigator turns out to be so poorly distributed that it no longer makes sense to consider it as such. In these cases, transforming the variable into one that is discrete and "binning" the variable often makes the most sense as a post hoc decision. As an example, suppose for instance that you distributed a survey to a sample in which you asked recipients how many times they got into a car accident in the past two years. On this variable, most individuals will likely respond with "0" times, while a few will respond with values of 1, and even less with values of 2,

3, etc. Consider how some fictitious data on such a distribution might look, where n_accidents is the number of car accidents an individual has experienced in the past two years:

```
> n_accidents <- c(0, 0, 0, 0, 0, 0, 0, 0, 0, 0, 0, 0, 0, 0,
0, 0, 0, 0, 0, 0, 0, 0, 0, 0, 0, 0, 1, 1, 1, 1, 1, 1, 1, 1,
1, 1, 1, 1, 1, 1, 2, 2, 2, 2, 2, 2, 3, 3, 3, 4, 4, 5)

> hist(n_accidents)
```

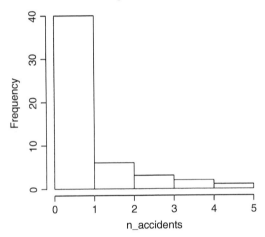

Histogram of n_accidents

The distribution of number of accidents is not only **noncontinuous**, but the values on this variable are also rather **poorly distributed**. Though there are statistical models that will accommodate such data, the issue in this case may not be so much about choosing a reasonable statistical model as it is about whether we are **measuring** something useful and meaningful. That is, if most of our sample is having zero car accidents, with hardly any individuals having greater than zero, the researcher in this case may find it more useful to simply transform the data into one that is **binary** in nature, perhaps binning all those individuals with zero car accidents into one category and "greater than 0" car accidents into another, generating the following distribution:

```
> n_accidents_bin <- c(0, 0, 0, 0, 0, 0, 0, 0, 0, 0, 0, 0,
0, 0, 0, 0, 0, 0, 0, 0, 0, 0, 0, 0, 1, 1, 1, 1, 1, 1,
1, 1, 1, 1, 1, 1, 1, 1, 1, 1, 1, 1, 1, 1, 1, 1, 1, 1, 1)
> hist(n_accidents_bin)
```

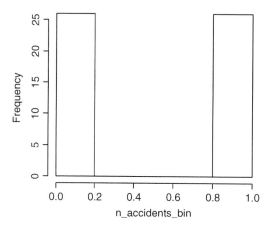

The variable **number of accidents** now has only two values (0 and 1), but R still believes it is continuous as evident by the scaling on the *x*-axis. We can ask R to shrink the scale down to two values:

```
hist(n_accidents_bin, xaxt = 'n')
axis(side=1, at=seq(0, 1))
```

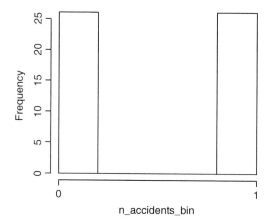

For variables such as this having a **binomial** (two categories) or **multinomial** (more than two categories) distribution, classic ANOVA or regression models are inappropriate, since these typically assume the response variable has sufficient distribution and could at least be argued to be continuous in nature. When the response variable is **binary** or **polytomous**, either naturally occurring or by design of the investigator as in the case of the number of car accidents, a new type of statistical model is required. These models, generally known as **generalized linear models**, effectuate a **transformation** on the binary or polytomous response such that the new transformed response can be deemed more or less continuous. It is on this transformed variable that the analysis then takes place. Though the class of generalized linear models is very large, in this chapter we will focus on one model primarily, in which the response variable is binary or multicategory. These models are known as **logistic regression models**. Other special cases of the generalized linear model include **Poisson regression** among others. For a full discussion of the generalized linear model, see McCullagh and Nelder (1990).

To help motivate our discussion of logistic regression models, we unpack a few fundamental concepts by beginning with a very simple example in R, and will extend on more of the theory and further applications in R a bit later. Hence, we begin very slowly to survey exactly how the logistic model differs from classic ordinary least-squares.

8.2 Example of Logistic Regression in R

We again consider fictional IQ data, where scores were obtained on quantitative and verbal ability in two groups. One group received training, while a second group did not. The group receiving training is coded "1" and the group not receiving training is coded "0." Our data appear in Table 8.1.

Table 8.1 Hypothetical data on quantitative and verbal ability for those receiving training (group = 1) versus those not receiving training (group = 0).

Subject	Quantitative	Verbal	Training group
1	5	2	0
2	2	1	0
3	6	3	0
4	9	7	0
5	8	9	0
6	7	8	1
7	9	8	1
8	10	10	1
9	10	9	1
10	9	8	1

We first generate this data in R, and confirm that it was entered correctly:

```
> q <- c(5, 2, 6, 9, 8, 7, 9, 10, 10, 9)
> v <- c(2, 1, 3, 7, 9, 8, 8, 10, 9, 8)
> train <- c(0, 0, 0, 0, 0, 1, 1, 1, 1, 1)
> qv.data <- data.frame(q, v, train)
> qv.data
      q   v  train
1     5   2      0
2     2   1      0
3     6   3      0
4     9   7      0
5     8   9      0
6     7   8      1
7     9   8      1
8    10  10      1
9    10   9      1
10    9   8      1
```

We note that `quantitative` and `verbal` are probably safe to consider as continuous variables, while `train` is definitely a binary variable. For this first example, suppose we are interested in **predicting train from knowledge of quantitative**. That is, if we know one's quantitative score, can we predict which train group they belong to? A glance at the data perhaps somewhat reveals that higher scores on quantitative overall seem to be linked to train = 1, while lower scores seem to be linked to train = 0.

Now, at first glance, this question looks a lot like a question we would ask in an ordinary regression framework. Indeed, had `train` been a continuous variable, and reasonably distributed, we could simply employ ordinary least-squares regression and be on our way. That is, we could designate `train` as our dependent variable, and quantitative as our predictor. However, since `train` is binary, a suitable strategy (among a few possible choices) is to employ the logistic regression model. We ask the question:

Is Train a Function of Quantitative?

Though there are assumptions for logistic regression as there are for most statistical models, we delay discussion of those for now and proceed directly to fitting the logistic model to these data:

```
> fit <- glm(train ~ q, data = qv.data, family=binomial())
> summary(fit)
```

The above glm() statement, which stands for **generalized linear model**, looks very similar to the traditional lm() statements we remember from regression. The family() option in the tail of the statement is where things are a bit different. This is where we specify the relevant **link function**, which is a subtype of the **glm** model. For our model, in specifying the binomial() family, we are invoking the **logit link**, which we will describe further very soon. The output now follows:

```
Call:
glm(formula = train ~ q, family = binomial(), data = qv.data)

Deviance Residuals:
    Min      1Q Median      3Q     Max
-1.6444 -0.4912 0.2091 0.7052 1.5668

Coefficients:
            Estimate Std. Error z value Pr(>|z|)
(Intercept)  -7.6466     5.2058  -1.469    0.142
q             0.9666     0.6220   1.554    0.120

(Dispersion parameter for binomial family taken to be 1)

    Null deviance: 13.863 on 9 degrees of freedom
Residual deviance:  8.745 on 8 degrees of freedom
AIC: 12.745

Number of Fisher Scoring iterations: 5
```

Again, we defer discussion of the complete output of logistic regression until later. For now, we focus only on the most essential components of the above analysis using it to introduce concepts fundamental to logistic regression. Inspecting the above output, we see that R generated the following coefficients for the model:

```
Coefficients:
            Estimate Std. Error z value Pr(>|z|)
(Intercept)  -7.6466     5.2058  -1.469    0.142
q             0.9666     0.6220   1.554    0.120
```

At first glance, these estimated coefficients look a lot like coefficients from a least-squares regression analysis. For now, we skip over interpreting the constant term, and instead move directly to interpreting the coefficient for q.

This value is equal to 0.9666. The fact that it is not statistically significant ($p = 0.120$) need not concern us right now, as we are more interested first and foremost in discussing what the number 0.9666 actually means. Recall that in a traditional least-squares regression, we would interpret the number as follows:

For a one-unit increase in Q, we would expect, on average, a 0.967 unit increase in the response variable.

In least-squares regression, the response variable mentioned above is, as already discussed, continuous. In the logistic regression we just ran, it is impossible to think of `train` as continuous since it only has two categories. Rather than being continuous, our response variable for this model is **binary** in nature. Hence, interpreting these coefficients as one would in an ordinary regression analysis will not work. To appreciate why, imagine the following least-squares interpretation for this data adapted from the above statement:

For a one-unit increase in Q, we would expect, on average, a 0.967 unit increase in train.

Now, ask yourself, on an intuitive level alone, does this interpretation sound correct? Does it make any sense? What would it mean for `train` to increase by 0.967 units? Train only has two values, so how can it increase 0.967 of a unit? Clearly, the logistic regression output is not operating on the **original variable** `train`. Rather, it is operating on a **transformed variable**. What we need to figure out now is the nature of this transformed variable and what 0.967 actually means so we can more accurately interpret the output from the logistic regression.

8.3 Introducing the Logit: The Log of the Odds

Logistic regression operates on a continuous variable similar to OLS regression, only that the continuous variable it operates on is a **transformed variable** and not the original binary variable. As discussed, we know it cannot be the original variable since the original variable has only two categories. In essence then, logistic regression seeks to transform this otherwise binary variable into something that is more or less continuous in nature. That new continuous variable we will call the **logit**, which as we will see is the **log of the odds**. Hence, the correct interpretation of our obtained coefficient for quantitative will be:

For a one-unit increase in Q, we would expect, on average, a 0.967 unit increase in the logit of the response.

Other than the logit being a continuous variable, we do not yet know anything about it. So what is it? Well, as mentioned, it's the **log of the odds**. But what does that mean? To understand this, we need to first understand what **odds** are. Odds can be very confusing, so we must be careful and cautious in our explanation of them. An example will help. Suppose we have an event that can or cannot occur. That is, it is **binary** in nature, and it is either **present** or **absent** (i.e. mutually exclusive events). One such example might be a symptom for a disease. Suppose the symptom is present or is absent. The odds of the event occurring are defined by:

$$\text{odds} = \left(\frac{p}{1-p}\right),$$

where p is the probability of the event **occurring**, and so $1 - p$ is the probability of the event **not occurring**. Notice that $1 - p$ is the balance of p, which implies that the event in question either occurs or does not occur. For instance, if we are speaking about a coin with two sides, heads versus tails, and p is the probability of heads, then $1 - p$ is the probability of tails. In this we assume the coin cannot land on its edge. If the probability of heads is 0.5, then p is equal to 0.5, and $1 - p$ is also equal to 0.5. Hence, the odds are $0.5/(1 - 0.5) = 1$. Odds of 1 mean the event occurring or not occurring is equally likely. You can think of the number 1 as being the "center" of fairness. Now, suppose you have a biased coin, and the probability of heads is 0.75. Then, the odds in favor of heads is $0.75/(1 - 0.75) = 0.75/0.25 = 3$. That is, the odds in favor of **heads to tails** are **3 to 1**. The odds against obtaining a head are 1 to 3. Notice we say "3 to 1" or "1 to 3" even though the only number we obtained was a 3. Since 1 is the "center," we always interpret the odds as "to 1."

So why introduce odds here? Odds are useful because, as mentioned, the response variable is binary, so in a binary logistic regression, odds are a useful way to think of the response. If 95 individuals out of 100 survive an invasive surgery, then the odds in favor of survival are $0.95/1 - 0.95 = 0.95/0.05 = 19$ to 1. The odds against survival are 1 to 19.

8.4 The Natural Log of the Odds

We have seen above how we can transform a probability into a statement about the odds in favor of an event. Odds result in whole numbers and in statements such as "3 to 1," "4 to 1," etc. That is, they aren't actually measurable on a continuous scale. However, in our logistic regression, we'd like to have something measurable on a continuous scale, and thus we would like to transform the odds into something that is continuous. The transformation will be to take the **natural log of the odds**, thereby transforming the odds into something that is approximately linear. That is, we will compute:

$$\ln(\text{odds}) = \ln\left(\frac{p}{1-p}\right).$$

Notice above that all we have done is compute the natural logarithm of the odds. This equation, $\ln\left(\frac{p}{1-p}\right)$, has a special name. It is called the **logit**. The logit is defined as the **log of the odds**, or sometimes also called the **logged odds**. For our coin example, recall the odds were 3 to 1. The probabilities were 0.75 to 0.25, which gave us 3. When we take the natural logarithm of 3, we get 1.0986. We can easily compute natural logarithms in R:

```
> log(3)
[1] 1.098612
```

This is the logit associated with an odds of 3 to 1. Likewise, the logit associated with odds of 1 to 3 (or 0.3333), that is, the probability 0.25, is −1.0987, as R readily tells us:

```
> log(0.3333)
[1] -1.098712
```

The key point to remember is this: **logits are transformed odds, and these logits are continuous as one would have in least-squares regression**. Hence, when we run the regression on these logits, we are, in a sense, doing something very similar to what we do in ordinary least-squares regression, since now the response variable, as it was in OLS regression, is continuous in nature. This is the key idea behind the generalized linear model and link functions in general, and that is to transform a noncontinuous and nonlinear response variable into something that is more similar to a scale we would operate on in ordinary regression. This is what generalized linear models are all about, using a link function to transform what is an otherwise nonlinear response variable into one that is approximately linear. Linear regression can be then interpreted in light of the generalized linear model in which the link function is that of the **identity function**. That is, since the response in a linear regression is already continuous and often approximately normally distributed, there is no transformation necessary. But to be more inclusive, we say the transformation is that of the identity function which means what you put in, you get out. So if you put in a value for x into the function, you obtain the same value. That is, $f(x) = x$ is the identity function.

8.5   From Logits Back to Odds

Given that we can convert odds to logits, we should then be able to convert logits back into odds. The inverse of the log function is the **exponential** function. For example, consider the logarithm of a number to base 2:

$$\log_2(8) = 3$$

The above means that 2 to the exponent 3 gives us 8. That is, the logarithm of 8 to base 2 is equal to 3. We see that the **logarithm, is in fact, an exponent** (sometimes simply called an "index" by which the base must be raised). The inverse of the logarithm is defined by "exponentiating" to base 2. This operation of taking the base 2, then raising it to 3 to get 8 is an example of exponentiating. In the case of the logit, since we have defined it as the natural logarithm of the odds, to get the original number back, we will exponentiate to base e, which is the base of a natural logarithm. Hence, in our example, had the base been e instead of 2, we would have:

$$\log_e(8) = 2.079$$

That is, $e^{2.079} = 8$. The number e is an **irrational number** and is a constant equal to approximately 2.71828. An irrational number in mathematics is a number that cannot be written as a quotient of two integers. It does not have a finite decimal point or one that is periodic. The constant e is irrational because when we compute it, it never seems to end and we do not notice any repetition in numbers such as we would have with a periodic decimal expansion (e.g. 4.567567567).

In our previous example, recall that the logit was equal to 1.0986. When we raise the number e to 1.0986, we obtain the odds:

$$e^{1.0986} = 2.71828^{1.0986}$$
$$= 3$$

Notice the number 3 is the same number we computed earlier for the odds. We can obtain the same number in R through the exp() function, which stands for the exponential function:

```
> exp(1.0986)
[1] 2.999963
```

Rounded, the actual number is 3. Hence, we can see that we can transform probability into odds, and odds into logits, then through exponentiating the logit, transform these back into odds and probabilities. The choice we make for interpretation often depends largely on which measure we prefer when reporting findings.

Logistic regression operates by first transforming the odds into the logged odds, or the "logit." Logits can then be converted back into statements about odds, as well as statements about probabilities. Logistic regression is thus similar in many respects to ordinary least-squares regression, only that it operates on a transformed response variable.

8.6 Full Example of Logistic Regression

8.6.1 Challenger O-ring Data

We now give a full example of fitting a logistic regression model in R, while simultaneously exploring the data and justifying why logistic regression is a better approach than traditional least-squares regression. We will also briefly discuss **assumptions of the model**. Recall in our earlier example we skimped a bit on interpreting all the output, and instead focused on the interpretation of the logit and the technical basis of logistic regression. Now, we are ready to perform a full analysis and full interpretation of output.

Our data come from a tragic event that occurred on January 28, 1986. Space shuttle Challenger lifted off from Cape Canaveral, Florida at 11:39 a.m. EST, and shortly after, exploded. The lives of all astronauts were lost in the event. It turned out that one of the causes of the accident was a failure of a seal on the shuttle's O-rings. These O-rings, when functioning properly, are designed to keep fuel inside the rocket booster and not allow it to leak out. The final investigation and report on the accident revealed that the O-ring likely failed as a result of the temperature at which Challenger was launched. The temperature that morning was 31°F, and very much colder than in any previous launch of the shuttle. Indeed, data from previous launches showed that potential failures were more likely to occur when temperatures were extremely low versus very high. Cold temperatures that morning apparently caused the O-ring to become dysfunctional, allowing fuel to leak out of the booster and onto the main fuel tank of the shuttle, which then presumably initiated the explosion.

The following are O-ring failure data based on previous launches that were collected prior to the launch of the shuttle. In this data, "1" is denoted as a failure (i.e. a problem with the O-ring), and "0" is denoted as "no failure" (i.e. no problem with the O-ring). Paired with each O-ring event (failure vs. no failure) is the associated temperature at which the event occurred:

```
> oring <- c(1, 1, 1, 1, 0, 0, 0, 0, 0, 0, 0, 0, 1, 1, 0, 0,
0, 1, 0, 0, 0, 0, 0)
> temp <- c(53, 57, 58, 63, 66, 67, 67, 67, 68, 69, 70, 70,
70, 70, 72, 73, 75, 75, 76, 76, 78, 79, 81)
> challenger <- data.frame(oring, temp)
> challenger
  oring temp
1    1 53
2    1 57
3    1 58
4    1 63
5    0 66
6    0 67
7    0 67
8    0 67
9    0 68
10 0 69
11 0 70
12 0 70
13 1 70
14 1 70
15 0 72
16 0 73
17 0 75
18 1 75
19 0 76
20 0 76
21 0 78
22 0 79
23 0 81
```

Let's first obtain a plot of these data:

```
> plot(temp, oring)
```

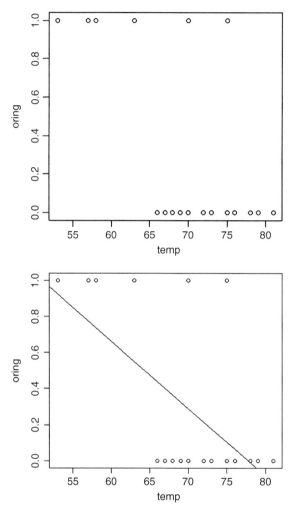

Notice that the data, graphed in the first plot, as we would expect, group at oring = 0 and oring = 1. That is, there is no distribution of oring values across the range of temp as one would typically have in an ordinary least-squares regression. As mentioned already, the fact that the data are **binary** on the response variable like this makes ordinary least-squares a poor option for analyzing these data. However, for demonstration only, let us first fit a linear model to these data to see what we get:

```
> challenger.linear <- lm(oring ~ temp)
> summary(challenger.linear)
```

```
Call:
lm(formula = oring ~ temp)

Residuals:
    Min      1Q   Median      3Q     Max
-0.43762 -0.30679 -0.06381 0.17452 0.89881

Coefficients:
            Estimate Std. Error t value Pr(>|t|)
(Intercept)  2.90476    0.84208   3.450  0.00240 **
temp        -0.03738    0.01205  -3.103  0.00538 **
---
Signif. codes:  0 '***' 0.001 '**' 0.01 '*' 0.05 '.' 0.1 ' ' 1

Residual standard error: 0.3987 on 21 degrees of freedom
Multiple R-squared: 0.3144,        Adjusted R-squared: 0.2818
F-statistic: 9.63 on 1 and 21 DF, p-value: 0.005383
```

The estimated linear regression equation for these data is thus

```
oring = 2.90476 - 0.03738*temp
```

If we fit a least-squares line to these data, we would obtain that given in the second plot above (i.e., with the fitted least-squares line). We generated the regression line using abline(challenger.linear). One of the problems with applying a linear regression approach to these data is that it can result in predicted values that make no sense. For instance, for a temperature of 85, the prediction on the response variable is:

```
oring = 2.90476 - 0.03738*temp
oring = 2.90476 - 0.03738*85
oring = -0.27254
```

But what does a predicted value for oring of −0.27254 represent? It's an illogical answer for the given data. A plot of residuals on these data also reveals nothing close to approximating a random distribution, as we would expect in least-squares. Recall that in verifying the assumptions of least-squares regression, we expected to see a relatively random distribution of predicted values when plotted against residuals. However, for the linear model we just fit to the challenger data, this is clearly not the case:

```
> fitted <- fitted(challenger.linear)
> resid <- residuals(challenger.linear)
> plot(fitted, resid)
```

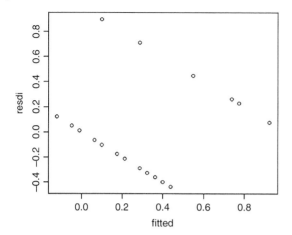

Clearly then, a linear model on these data may not be ideal. We need a new approach, which is why we will turn to logistic regression to analyze these data.

8.7   Logistic Regression on Challenger Data

Having tried a traditional linear model and having found it problematic, we now fit the correct model. Using a logistic regression, we model `oring` as a function of `temp`:

```
> challenger.fit <- glm(oring ~ temp, data = challenger,
family = binomial())
> summary(challenger.fit)

Call:
glm(formula = oring ~ temp, family = binomial(), data = challenger)

Deviance Residuals:
    Min      1Q   Median      3Q     Max
-1.0611 -0.7613 -0.3783 0.4524 2.2175

Coefficients:
            Estimate Std. Error z value Pr(>|z|)
(Intercept)  15.0429     7.3786   2.039   0.0415 *
temp         -0.2322     0.1082  -2.145   0.0320 *
---
Signif. codes: 0 '***' 0.001 '**' 0.01 '*' 0.05 '.' 0.1 ' ' 1
```

```
(Dispersion parameter for binomial family taken to be 1)
    Null deviance: 28.267 on 22 degrees of freedom
Residual deviance: 20.315 on 21 degrees of freedom
AIC: 24.315
Number of Fisher Scoring iterations: 5
```

As before, the last part of the call `family = binomial()` requests the "logit" link, since this is the default used by R. Hence, we could have also coded `family = binomial("logit")` and obtained the same output (Dalgaard, 2008).

Let's unpack each part of the output:

- The `null deviance` is the deviance of a model containing only the intercept in the model. The exact meaning of deviance here is difficult to explain without reference to more theory. For our purposes, a difference between the residual deviance and the null deviance will help us know whether our predictor variable (`temp`) is useful. Ideally, we would want residual deviance to be less than the null deviance, as such would signify a reduction in "error" (to use the least-squares analogy).
- The `residual deviance` is analogous to residual sums of squares in OLS regression, which in OLS corresponds to an estimate of variation around the fitted regression line. As mentioned, the residual deviance is computed after bringing in `temp` as a predictor into the model.
- As mentioned above, if the model is "successful," then we would expect a **drop in deviance** from the null to residual. In the current case, the drop in deviance from 28.267 to 20.315 suggests that `temp` may be useful as a predictor.
- The value of AIC is equal to 24.315. Recall that lower values of AIC are preferable to larger values, and that AIC is especially useful for comparing models.
- `Fisher scoring iterations`: 5 refers to how many times the underlying algorithm had to iterate on the parameter estimates before a stable solution was found. For applied purposes, this number need not concern us.

8.8 Analysis of Deviance Table

Above, we alluded to the concept of **deviance** in the output. We can also request in R what is known as an **analysis of deviance** table:

```
> anova(challenger.fit, test="Chisq")
Analysis of Deviance Table
Model: binomial, link: logit

Response: oring

Terms added sequentially (first to last)
```

```
        Df Deviance Resid.  Df Resid. Dev Pr(>Chi)
NULL                       22      28.267
temp    1   7.952          21      20.315 0.004804 **
---
Signif. codes: 0 '***' 0.001 '**' 0.01 '*' 0.05 '.' 0.1 ' ' 1
```

We see from the table the same information that the primary output featured earlier under `Resid.Dev`, which is the drop in deviance from the null model to the model that includes `temp` (i.e. from 28.267 to 20.315). The difference in deviance is given under the column **Deviance**, equal to 7.952 (i.e. $28.267 - 20.315 = 7.952$). The differences in models can be tested via a chi-squared test, yielding a *p*-value of 0.004804. Hence, we see that by including the predictor `temp`, a statistically significant drop in deviance occurs. That is, the *p*-value of 0.004804 is telling us that the drop in deviance from 28.267 to 20.315 is statistically significant, suggesting that including `temp` as a predictor reduces the error in the model more than would be expected by chance.

We can obtain **confidence intervals** for both the intercept and predictor via `confint.default()`:

```
> confint.default(challenger.fit)
                 2.5 %        97.5 %
(Intercept)  0.5810523  29.50475096
temp        -0.4443022  -0.02002324
```

8.9   Predicting Probabilities

Now that we've fit our logistic regression model, we would like to use it to predict probabilities on new data. Suppose, for instance, that NASA engineers would like to predict the probability of failure for a launch when temperature is equal to 30. To obtain these, we will create a new data frame called `predict.prob`, and then request predicted values. We compute:

```
> predict.prob <- data.frame(temp = 30)
> predict(challenger.fit, type="response", newdata =
predict.prob)
 1
0.9996898
```

For a temperature of 30 then, the predicted probability of failure is equal to a whopping 0.9996898. Recall that Challenger was launched at a temperature of 31.8°F. How about for a temperature of 90°F?

```
> predict.prob <- data.frame(temp = 90)
> predict(challenger.fit, type="response", newdata =
predict.prob)
 1
0.002866636
```

We see that for a relatively high temperature, the probability of failure is quite small.

We next obtain predicted logits for cases in our data 1 through 23:

```
> predict(challenger.fit)
         1          2          3          4          5
 2.7382762  1.8096252  1.5774625 0.4166488 -0.2798395
         6          7          8
-0.5120022 -0.5120022 -0.5120022
         9         10         11         12         13
-0.7441650 -0.9763277 -1.2084904 -1.2084904 -1.2084904
        14         15         16
-1.2084904 -1.6728159 -1.9049787
        17         18         19         20         21
-2.3693042 -2.3693042 -2.6014669 -2.6014669 -3.0657924
        22         23
-3.2979551 -3.7622806
```

We see from the above that the predicted logit for the first observation in the data is 2.7382762, followed by 1.8096252 for the second, and so on. To demonstrate the "long way" how the above were computed, let's compute them for ourselves using the original estimated model equation:

```
> y <- 15.0429 -0.2322*temp
> y
 [1]  2.7363 1.8075 1.5753 0.4143 -0.2823 -0.5145 -0.5145
-0.5145 -0.7467
[10] -0.9789 -1.2111 -1.2111 -1.2111 -1.2111 -1.6755
-1.9077 -2.3721 -2.3721
[19] -2.6043 -2.6043 -3.0687 -3.3009 -3.7653
```

We see that within rounding error, our computed logits match those generated by R.

8.10 Assumptions of Logistic Regression

Recall that the assumptions of ordinary least-squares regression included independence and normality of errors, as well as homoscedasticity of errors. It was also assumed that collinearity among predictors was absent, which could be verified via the **variance inflation factor** (**VIF**). The assumption of independence carries over to logistic regression. That is, errors in logistic regression are assumed to be independent (Stoltzfus, 2011; Tabachnick and Fidell, 2001), which implies that a given error should not be predictive of any other error. Of course, this assumption will be violated with repeated-measures data or any other data that has correlated outcomes, whether in OLS regression or logistic regression. Recall that the assumption of independence is usually more or less assured by the type of data one collects, as well as the method of data collection, though at times residual plots can potentially prove useful in attempting to detect anomalies in this regard (though usually insufficient on their own for confirming a lack of independence, still, they can be helpful). As was the case for OLS regression, Cook's d can also be used to detect influential values in logistic regression. Residual plots can also be obtained (see Agresti, 2002; Fox, 2016).

So far then, the assumptions for logistic regression more or less parallel those for OLS regression. However, recall the assumption of linearity. In OLS regression, it was assumed that the relationship between response variable and predictor(s) was more or less linear in form. Even if it wasn't exactly, we still had to assume linearity in the parameters. Residual plots were helpful in evaluating this assumption. In logistic regression, we are likewise needing to verify linearity, but now, since the response variable is binary, the linearity we need to satisfy is between continuous predictor(s) and the logit, rather than the actual values of the response variable, which recall for logistic regression have binary values. There are a few different ways of assessing this assumption, though a popular option is the Box-Tidwell approach (Hosmer and Lemeshow, 1989). Typically, the assumption will more or less be satisfied, or at minimum not violated "enough" so that we cannot proceed with the logistic regression or need to transform variables. Hence, we do not demonstrate the test here. For details on the test in general, see Tabachnick and Fidell (2001).

8.11 Multiple Logistic Regression

Analogous to the case where we can model several predictors in multiple linear regression, **multiple logistic regression** also allows us to include several predictors. The concept is the same, though as was true for the one-predictor case, the interpretation of parameter estimates will be different in multiple logistic regression than in simple. However, the idea of "context" will still apply. That is, the effect any predictor on the logit must be interpreted within the context of what other variables have also been included in the regression.

Let's consider an example. Recall our q.v. data featured earlier, in which we only used quantitative ability as a predictor. Suppose now we wish to use verbal simultaneously in a model with quantitative:

```
> fit.qv <- glm(train ~ q + v, data = qv.data,
family=binomial())
> summary(fit.qv)

Call:
glm(formula = train ~ q + v, family = binomial(), data = qv.data)

Deviance Residuals:
    Min      1Q   Median      3Q      Max
-1.7370 -0.1247  0.1507  0.7686  1.1691

Coefficients:
            Estimate Std. Error z value Pr(>|z|)
(Intercept)  -9.4991     9.8068  -0.969    0.333
q             0.3919     0.9332   0.420    0.674
v             0.8469     0.9901   0.855    0.392

(Dispersion parameter for binomial family taken to be 1)

    Null deviance: 13.8629 on 9 degrees of freedom
Residual deviance:  7.5858 on 7 degrees of freedom
AIC: 13.586

Number of Fisher Scoring iterations: 7
```

With both predictors simultaneously in the model, our interpretation of coefficients is now a bit different than in the case of only a single predictor because we must take into account the presence of the second predictor. We disregard statistical significance for now, and simply demonstrate the interpretation of the coefficients:

- For a one-unit increase in quantitative ability, we expect, on average, a 0.3919 increase in the logit given that verbal ability is also simultaneously included into the model.
- For a one-unit increase in verbal ability, we expect, on average, a 0.847 increase in the logit given that quantitative ability is also simultaneously included into the model.

Again, be careful to note that including the qualifying statements of "given that verbal/quantitative is also simultaneously included in the model" is key here and very important. Without including this "context," the interpretation of coefficients is, strictly speaking, incorrect.

We can compute odds ratios (Cohen et al., 2003) for these predictors:

```
> exp(coef(fit.qv))
(Intercept)          q                    v
0.000074921    1.479851678          2.332398240
```

We interpret the above odds ratios:

- For a one-unit increase in quantitative ability, the odds are approximately 1.48 of being in group 1 versus 0, given the simultaneous inclusion of verbal in the model.
- For a one-unit increase in verbal ability, the odds are approximately 2.33 of being in group 1 versus 0, given the simultaneous inclusion of quantitative into the model.

Analogous to interpreting logits, because the above odds ratios for each predictor occur in a model with other predictors simultaneously also in the model, and hence are "adjusted" for the inclusion of the other predictors, they are sometimes referred to as **adjusted odds** or **adjusted odds ratios** in this context (for a discussion of odds ratios, see Cohen et al. 2002). This is an analogous concept to least-squares coefficients being "adjusted" for other predictors simultaneously in the model in multiple regression. **Remember that any parameter estimate in a model is always interpreted relative to the model and not independent of it. No parameters are estimated in a vacuum, even if the model is extremely simple. All statistical models are context-dependent, which means parameter estimates in any model depend on what other predictors are included (or not included).**

We can obtain the probabilities of being in the training group via `predict()`. We request only for cases 1 through 5 using `[1:5]`:

```
> glm.probs = predict(fit.qv, type = "response")
> glm.probs[1:5]
            1            2            3            4
0.0028843490 0.0003825417 0.0098857242 0.4891729541
            5
0.7787741975
```

Next, we obtain a **confusion matrix** containing correct and incorrect classifications:

```
   FALSE  TRUE
0    4     1
1    0     5
```

How well did our logistic regression model do? We can see from the above that it correctly classified those in the training group a total of 5 times, and those with no training a total of 4 times. There was one **error in classification**, represented in the top right of the confusion matrix. We can compute the total proportion of cases correctly classified as the sum of cases along the main diagonal divided by the total number of cases. That is, the total proportion of correctly classified cases is:

```
> (4 + 5)/10
[1] 0.9
```

 When performing a multiple logistic regression, as was true for multiple regression, the caveat "given the model" or "given the inclusion of the other predictors" or "holding other predictors constant" is necessary. That is, interpreting coefficients in a logistic regression model as though they were not adjusted for other predictors is incorrect.

8.12 Training Error Rate Versus Test Error Rate

We computed above that 90% of cases were correctly classified. Hence, the error rate is equal to the balance of this, which is $100 - 90 = 10\%$. That is, for our data, 10% of cases were incorrectly classified. However, it should be emphasized that this computed rate refers to what is known as the **training error rate**, because it is computed on the original data set, which is often referred to as the **training data**. This error rate is also known as the **apparent error rate**. In many cases, the training data may be the only data the researcher has available, especially if the data are somewhat small as in the current situation. The **test error rate** is the rate computed on a cross-validated sample. That is, this is the error rate that occurs when we take the obtained model and use it to predict responses on another data set, presumably sampled from the same population as the original data set. The error rate computed on this new sample can be used to estimate the **actual error rate**.

So why does all this matter? You can probably already foresee the problem. Since the training error rate is computed on the same data used to derive the model, this computed error rate will be overly **optimistic**, meaning that the actual test error rate is typically expected to be greater (Rencher and Christensen, 2012). Ways of obtaining a better estimate of the true error rate include splitting the sample into two, running the model on the first part, then validating on the second part. However, this typically requires a relatively large sample size. See Rencher and Christensen (2012) for details, and also Hastie et al. (2009) for an even more thorough discussion of training versus test error rates.

As was true for linear regression, stepwise models in logistic regression can also be fit using the `stepAIC()` function in the **MASS** package. A competitive alternative to logistic regression as well as discriminant analysis is the machine learning technique of **support vector machines**. For details, see Izenman (2008).

Exercises

1 Discuss the general purpose of logistic regression, and how it is similar and different from that of ordinary least-squares regression.

2 Give two examples of research scenarios where logistic regression would be advised over least-squares regression. For the first, make the response variable naturally occurring binary. For the second, assume the investigator initially assumed the response would be continuous and then saw the need to transform it into a binary variable.

3 Under what circumstance is a coefficient for logistic regression interpreted in a similar way as that in ordinary least-squares regression?

4 Why is it necessary to take the log of the odds in logistic regression? Why not simply perform the logistic regression directly on the odds?

5 Referring to the equation for the odds, for what value of p would yield a logit of 0? What does this mean then for logits of value less than 0?

6 What does it mean to "exponentiate the logit?" Give a simple example of such a transformation.

7 In the Challenger example in the chapter, the coefficient for temp was equal to -0.2322 with associated p-value equal to 0.0320. Interpret. What was the null hypothesis under test here?

8 Recall the achievement data (Table 8.2). Treat the textbook variable as the response variable with binary coding (code textbook 2 as "1" and textbook 1 as "0"), and use achievement to predict textbook used. Conduct the full logistic regression, and provide a general report and interpretation of findings.

Table 8.2 Achievement as a function of teacher and textbook.

	Teacher			
Textbook	**1**	**2**	**3**	**4**
1	70	69	85	95
1	67	68	86	94
1	65	70	85	89
2	75	76	76	94
2	76	77	75	93
2	73	75	73	91

9 For the analysis in #8, obtain an analysis of deviance table, and confidence intervals for the odds ratio. Interpret both.

10 For the analysis in #8, obtain a predicted probability for a student with an achievement score of 100.

9

Multivariate Analysis of Variance (MANOVA) and Discriminant Analysis

---

LEARNING OBJECTIVES

- Understand how multivariate analysis of variance (MANOVA) differs from univariate ANOVA.
- Understand when MANOVA is an appropriate model, and when not to use it.
- Understand how tests of significance differ in MANOVA compared to ANOVA, and why in the former there are several different kinds of tests.
- Understand how to compute effect size for MANOVA.
- Understand how to compute the Box-M test to evaluate the equality of covariance matrix assumption.
- Understand the nature of linear discriminant function analysis, and why it is sometimes used as a follow-up to MANOVA.
- Understand how to compute discriminant function scores, and use them to predict group membership on the response variable.
- Understand how to generate plots to visualize separation from discriminant functions.

Recall that in previous chapters, we introduced and developed the analysis of variance (ANOVA) model. ANOVA was useful for situations in which we had a single continuous dependent variable and one or more categorical independent variables called "factors." We referred to ANOVA as univariate because it consisted of only a single dependent variable, and called it one-way ANOVA if it had only a single independent variable. Factorial ANOVAs still had a single dependent variable, but included several independent variables all considered simultaneously, which allowed us to model potential interactions among these independent variables.

Modeling a single dependent variable at a time is suitable for many research situations. For example, recall that we hypothesized whether achievement was a function of teacher. The variable of achievement stood alone as the sole dependent variable in the model. However, there are times when we would like to model **two or more dependent variables simultaneously**. For instance, suppose we

Univariate, Bivariate, and Multivariate Statistics Using R: Quantitative Tools for Data Analysis and Data Science, First Edition. Daniel J. Denis.

wished to evaluate mean differences on variables verbal, analytic, and quant, considered simultaneously. That is, our new dependent or response variable is now a **combination** of these three variables of verbal, analytic, and quant. The type of combination will be a weighted sum of **verbal + analytic + quant**, which we will refer to as a **linear combination** of dependent variables. Then, as we did in ANOVA, we will be able to hypothesize this linear combination as a function of one or more independent variables. Should we find evidence of mean differences in such a model, it will be on the linear combination of dependent variables, and not each dependent variable considered separately. There are good reasons why we may wish to consider a response variable as a linear combination such as this. We discuss such reasons shortly.

Should we find evidence for mean differences on this linear combination of dependent variables, we may wish to study this linear combination in more detail. We may ask the question, "What is the make-up of the linear combination of dependent variables that maximizes separation between groups on the independent variable?" It turns out there will usually be more than a single linear combination that accomplishes the separation. These linear combinations that maximize separation among groups are called **discriminant functions**, and the study of such functions is useful as follow-ups to conducting a MANOVA. This area of study goes by the name of **linear discriminant function analysis**.

By the end of the chapter, we will see that both MANOVA and discriminant analysis are essentially, at a technical level, **opposites** of one another. We begin our discussion then with an introduction to MANOVA, and survey some of its technical basis before demonstrating the technique in R. We then follow up with a discussion and demonstration of discriminant function analysis.

9.1 Why Conduct MANOVA?

The first question to ask when considering a multivariate analysis of variance (or any other statistical procedure for that matter) is **whether you actually want to perform one**. Having consulted in statistics for a number of years now, I can tell you that for many research problems, univariate ANOVAs are often more appropriate for a researcher's data than a multivariate model. Some authors on applied statistics (e.g. see Everitt and Hothorn, 2011) have even forgone discussing MANOVA altogether, considering it an outdated technique given that in many instances, even after performing a MANOVA, researchers will nonetheless obtain univariate results on each dependent variable separately. However, we see that MANOVA is still useful both as a statistical method in its own right, and even more, is pedagogically meaningful as a stepping stone to understanding other techniques such as discriminant analysis and even the conceptual foundations of factor analysis incorporating ideas of **latent variables** and **constructs**. The point we wish to make is that the existence of a more sophisticated method of analysis does not necessarily imply it should be used. **Complexity in statistical method is not justification alone for using it**. From a scientific perspective, simpler is usually better, yet as Einstein once reminded us, no simpler than that.

So when does MANOVA make sense to perform? Though as will see there are statistical "perks" and advantages to doing MANOVA over several univariate ANOVAs, the bottom-line justification for doing MANOVA in general should be a **theoretical** or **research argument** that the dependent variables you have available actually should be considered simultaneously as a **linear combination**. For instance, it may make sense to conduct a MANOVA on the linear combination of verbal + analytic + quant, since the consideration of these variables simultaneously makes good substantive sense. That is, together, they can be argued to form what is known in general as a **variate**, a combination of variables. The name of the variate might be **IQ**, in this case. The key point here is that the consideration of these variables together often only makes good scientific or research sense if their combination is a **plausible construct**. Had our hypothesized linear combination been that of verbal + analytic + favorite ice cream, though a MANOVA is mathematically do-able on such a combination (i.e. the mathematics really don't have the ability to monitor what they are being applied to), we would nonetheless not do it since the combination of variables does not make any theoretical sense to us.

There is a second reason to sometimes prefer MANOVA over several independent ANOVAs, and that reason is purely **statistical** in nature as opposed to substantive. Recall that with every statistical test, with every rejection of a null hypothesis, there is a risk of a type I error, often set at $\alpha = 0.05$ or similar for the given test. As we conduct more and more tests on the same data, this error rate compounds and grows larger. For example, for three tests, each at $\alpha = 0.05$, the error rate is roughly $0.05 + 0.05 + 0.05 = 0.15$. It turns out that it isn't exactly equal to a simple sum (it actually turns out to be approximately 0.14 in this case, see Denis (2016, p. 485) for details), but the point is that when we conduct several univariate tests on the same data, the type I error rate grows larger and larger. Why is this a problem? It's a problem because usually we like to keep the type I error rate at a nominal level and have some degree of control over it. If the type I error rate gets too high, then we risk rejecting null hypotheses when in fact those null hypotheses may not actually be false. To refresh our memory here, recall what setting $\alpha = 0.05$ means for a given test. It means the probability of falsely rejecting the null is equal to 0.05, which scientifically, means that we might conclude a mean difference when in fact such a difference does not actually exist in the population from which the data were drawn. The bottom line is that we do not want our type

I error rate to unduly inflate, and when we perform a single MANOVA instead of several independent ANOVAs, we exercise control over the type I error rate, keeping it at a nominal level.

To reiterate, however, a **MANOVA should not be done simply to control type I error rates**. The first and foremost reason for doing a MANOVA should be because on a scientific basis, it makes the most sense. That is, if combining dependent variables into a single analysis is scientifically reasonable for your project, then MANOVA is an option. That it keeps type I error rates at a nominal level is an advantageous "perk," but it cannot be your sole reason for preferring MANOVA over several independent ANOVA models. That is, if substantively performing a MANOVA makes little sense, then it should not be performed regardless of its control over type I error rates.

The primary reason for conducting a MANOVA on a set of dependent variables considered simultaneously rather than several univariate ANOVAs on each dependent variable requires a research or logical justification that the dependent variables you have in your model go together well in some substantive sense. Hence, the first and foremost concern is whether it is reasonable to consider the dependent variables together as a set. A second reason to prefer MANOVA is that it helps mitigate the overall type I error rate. However, this second reason should never dominate the first. If your MANOVA is illogical, and combining dependent variables doesn't make good substantive sense, then benefitting from minimizing the type I error rate is not justification alone for performing the MANOVA over several univariate ANOVAs.

9.2 Multivariate Tests of Significance

Because of the fact that MANOVA tests all dependent variables simultaneously in a single model, the multivariate analysis of variance is in itself a different type of animal than univariate ANOVA. It is much more complex in its configuration, and how the dependent variables relate to one another in determining multivariate statistical significance is much more intricate than in the much simpler univariate ANOVA case. Recall that in ANOVA, the test of the so-called omnibus null hypothesis $H_0 : \mu_1 = \mu_2 = \mu_3 \cdots = \mu_j$ took the form of an F statistic,

$$F = \frac{\text{MS between}}{\text{MS within}},$$

which featured a comparison of MS between to MS within, and was evaluated on an F sampling distribution having $J - 1$ (numerator) and $N - J$ (denominator) degrees of freedom, where N was total sample size, and J the number of groups on the independent variable. This test served us well as an overall test in the univariate landscape because we had only a single dependent variable in the ANOVA. However, in the multivariate landscape, things have become a lot more complex as a result of modeling several dependent variables simultaneously. There is no single universal test of a multivariate hypothesis as there is in the ANOVA setting, and several competing multivariate tests have been historically proposed to evaluate multivariate hypotheses.

Why should we need different tests in a multivariate context? Why isn't the good 'ol F ratio sufficient? The primary reason why the F statistic is insufficient in a multivariate context is that we now have **covariances among dependent variables**, and these covariances must be included in any test of significance of a multivariate hypothesis. That is, any "good" test of a multivariate hypothesis should incorporate covariances among dependent variables since we are modeling these dependent variables now simultaneously. In the ANOVA context, we did not have any covariances among dependent variables since we only had a single response. Hence, one reason why multivariate models are, by their nature, more complex than univariate models, is that they must incorporate the degrees of covariance among responses, and because of that, there is no universal test of a null hypothesis in MANOVA. Rather, there are competing tests, each with differing ways of modeling that covariance, and resulting in different degrees of statistical power. What is more, finding a multivariate effect does not necessarily imply univariate effects on the dependent variables making up the multivariate one. And, vice versa, univariate significance does not imply multivariate significance. This is generally known as **Rao's paradox**, and is one reason why following up a MANOVA with univariate ANOVAs on each response variable, while do-able, is a **test of an entirely different set of hypotheses**, rather than a true "partition" of the multivariate effect.

A deeper discussion of the various multivariate tests requires knowledge of matrix and linear algebra. We surveyed elements of these topics earlier in the book, but still at a relatively elementary level. For deeper matrix treatments of these tests, see Johnson and Wichern (2007). Since the current book is primarily focused on applications of statistics with an emphasis on R software, instead of going into great detail into the matrix arguments and relationships among tests, we provide only a summary of these tests treading relatively light on the technical details. We spend the majority of our time fitting and interpreting multivariate models in R. A summary of the various multivariate tests now follows. Still, we can't escape a bit of technical detail with regard to these tests, as we will see.

1) **Wilks' Lambda**, $\Lambda = \dfrac{|\mathbf{E}|}{|\mathbf{H} + \mathbf{E}|}$. Wilks is what is known as an **inverse criterion**, which means that smaller values rather than large values count as evidence against the multivariate null hypothesis. Recall that in the traditional *F*-ratio of ANOVA, larger values than not counted against the null. In Λ, however, it stands to reason that evidence against the null should occur when $\mathbf{H}$ is much larger than $\mathbf{E}$, where $\mathbf{H}$ is the **hypothesis** matrix, and $\mathbf{E}$ the **error** matrix. Take a look at Λ and consider the extremes. That is, suppose $\mathbf{H}$ is a number greater than 0, and $\mathbf{E}$ is equal to 0. Under such a case, we would have

$$\Lambda = \frac{|\mathbf{E}|}{|\mathbf{H} + \mathbf{E}|} = \frac{0}{|\mathbf{H} + 0|} = \frac{0}{|\mathbf{H}|} = 0.$$

That is, if $\mathbf{H}$ were accounting for all the variability, then Λ would equal 0. What if instead $\mathbf{E}$ were accounting for all the variability, and $\mathbf{H}$ none? Then it stands that

$$\Lambda = \frac{|\mathbf{E}|}{|\mathbf{H} + \mathbf{E}|} = \frac{|\mathbf{E}|}{|0 + \mathbf{E}|} = \frac{|\mathbf{E}|}{|\mathbf{E}|} = 1.$$

Notice then that when $\mathbf{E}$ dominates relative to $\mathbf{H}$, we expect values increasingly larger, of which the maximum Λ can attain is 1. The researcher, however, is "cheering for" $\mathbf{H}$, which means smaller values than not are indicative of a potential effect. Wilks' lambda is a popular test used to evaluate multivariate effects, has a deep history, and is traditionally reported by most software. It is often the first test analysts will look at to see if there is evidence of a multivariate effect.

2) **Pillai's Trace**, $V^{(s)} = \mathrm{tr}[(\mathbf{E} + \mathbf{H})^{-1}\mathbf{H}]$. The notation "tr" means to take the trace of $(\mathbf{E} + \mathbf{H})^{-1}\mathbf{H}$. Since $\mathbf{E} + \mathbf{H} = \mathbf{T}$, we can rewrite $(\mathbf{E} + \mathbf{H})^{-1}\mathbf{H}$ as $\mathbf{T}^{-1}(\mathbf{H})$. So we see that Pillai's is taking the trace of the matrix $\mathbf{H}$ relative to $\mathbf{T}$, and where in Wilks we wanted the value to be smaller rather than larger, in Pillai's, large values are indicative of evidence against the multivariate null hypothesis. Pillai's can also be written as a function of **eigenvalues**, that is, $V^{(s)} = \displaystyle\sum_{i=1}^{s} \frac{\lambda_i}{1 + \lambda_i}$. We will use this interpretation a bit more once we arrive at discriminant analysis.

3) **Roy's Largest Root**: $\theta = \dfrac{\lambda_1}{1 + \lambda_1}$. Here, λ_1 is the largest of the eigenvalues extracted. Contrary to Pillai's, Roy's does not sum the eigenvalues as in $\displaystyle\sum_{i=1}^{s} \frac{\lambda_i}{1 + \lambda_i}$. Roy's only uses the largest of the extracted eigenvalues.

4) **Lawley–Hotelling's Trace:** $U^{(s)} = \text{tr}(\mathbf{E}^{-1}\mathbf{H}) = \sum_{i=1}^{s} \lambda_i$. Here, $U^{(s)}$ takes the trace of $\mathbf{H}$ to $\mathbf{E}$.

In most situations, all of the above multivariate tests will suggest the same rejection or nonrejection on the null hypothesis. In cases where they do not, you are strongly advised to check in with a statistical consultant to explore why they are suggesting different decisions on the null. Patterns of covariances among dependent variables may be at play, and a consultant may require you to run further analyses in an attempt to explore or discover such patterns. When tests do suggest different decisions on the null, it may be due to differing levels of statistical power and how mean vectors are aligned in multivariate space. For more details on the different tests, consult Olson (1976) or any good book on multivariate analysis that discusses MANOVA in some depth. The aforementioned Johnson and Wichern (2007) is one such book and is a classic reference on multivariate models in general. Rencher and Christensen (2012) has also become a classic and features an extensive discussion of MANOVA models.

9.3 Example of MANOVA in R

We now demonstrate MANOVA in R. We again consider data on IQ where participants were randomly assigned to one of three training groups. In group 1, no training was received. In group 2, some training was received, and in group 3, much training was received. Be careful to note the hypothesis we are interested in, as it is different from a hypothesis on each dependent variable considered univariately. In the MANOVA, we are interested in learning whether taken as a **set**, and **considered simultaneously**, there might be mean vector differences between training groups on a linear combination of these dependent variables. In R, we first generate the data frame:

```
> quant <- c(5, 2, 6, 9, 8, 7, 9, 10, 10)
> verbal <- c(2, 1, 3, 7, 9, 8, 8, 10, 9)
> train <- c(1, 1, 1, 2, 2, 2, 3, 3, 3)
> iq.train <- data.frame(quant, verbal, train)
> iq.train
  quant verbal train
1     5      2     1
2     2      1     1
3     6      3     1
4     9      7     2
5     8      9     2
```

6	7	8	2
7	9	8	3
8	10	10	3
9	10	9	3

Notice that our data frame has a total of 9 observations, 3 in each training group. Our dependent variables are quant and verbal. Since it is a MANOVA, we want to analyze these responses simultaneously. Hence, we first need to generate the new dependent variable, which is the combination of quant and verbal. We will call this new variable by y:

```
> y <- cbind(quant, verbal)
> y
     quant verbal
[1,]     5      2
[2,]     2      1
[3,]     6      3
[4,]     9      7
[5,]     8      9
[6,]     7      8
[7,]     9      8
[8,]    10     10
[9,]    10      9
```

Next, we designate train as a factor, and give the levels of the factor names "none," "some," and "much," where 1:3 denotes the three levels:

```
> train.f <- factor(train, levels = 1:3)
> levels(train.f) <- c("none", "some", "much")
> train.f
[1] none none none some some some much much much
Levels: none some much
```

We will need to discuss and verify assumptions for the MANOVA, but for now let's run the MANOVA so we can get an idea of what the output looks like. We hypothesize y as a function of train.f:

```
> manova.fit <- manova(y ~ train.f)
> summary(manova.fit)

          Df Pillai approx F num Df den Df  Pr(>F)
train.f    2 1.0737   3.4775      4     12 0.04166 *
Residuals  6
---
Signif. codes:  0 '***' 0.001 '**' 0.01 '*' 0.05 '.' 0.1 ' ' 1
```

By default, R reports **Pillai's trace**. We see that it is statistically significant ($p = 0.04166$, which is less than a conventional level set at 0.05) indicating that there are mean vector differences between train groups on the linear combination of quant and verbal. Again, recall what this means; in univariate ANOVA, a rejection of the null suggested there were mean differences on a single dependent variable. In MANOVA, since we are analyzing more than a single response variable, the rejection of the null is on a **linear combination of these variables**, not on any single variable considered univariately such as in classic ANOVA. This distinction is important. Again, the rejection of the null in this case means that there is evidence of a mean vector difference in training groups on quant and verbal, or, equivalently, a difference on a linear combination of quant and verbal.

As noted above, R generated Pillai's by default. However, we know based on our previous discussion that there is more than a single test available in MANOVA. Hence, we can request other tests as well. For instance, we can ask R to produce classic **Wilks' lambda**:

```
> summary(manova.fit, test = "Wilks")
          Df   Wilks approx F num Df den Df    Pr(>F)
train.f    2 0.056095   8.0555      4     10 0.003589 **
Residuals  6

- - -
Signif. codes:  0 '***' 0.001 '**' 0.01 '*' 0.05 '.' 0.1 ' ' 1
```

Though Wilks' lambda is statistically significant, $p = 0.003589$, we note that the p-value is smaller than it was for Pillai. This isn't by accident. Recall we mentioned that multivariate tests have differing levels of statistical power, and hence we would expect at times p-values to vary among tests. In both cases, however, the obtained p-values still suggest a rejection of the null hypothesis. Also, recall that from a scientific perspective, as opposed to statistical, a p-value that creeps up just above some designated significance level is not necessarily cause to conclude no evidence was found. For instance, had we obtained $p = 0.06$ or 0.07, though that doesn't meet the standard of 0.05, it certainly does not mean we should throw out our analysis and conclude nothing was found. The reason for the slightly higher p-value could, after all, be due to insufficient power (had the experiment not been properly planned to ensure adequate sample size). As always, look to **effect size** to know if something did or did not happen in the experiment or study. We consider effect sizes for MANOVA next.

9.4 Effect Size for MANOVA

Recall that when reporting any research finding, an estimate of the effect size should also be provided along with the significance test. Recall the reason for this. While a p-value indicates inferential support, an effect size gives you an idea of the scientific finding or degree of association, if there is one. And while we know based on our previous discussion that p-values can largely be a function of sample size, effect sizes are much more stable as sample size increases or decreases. Now that doesn't mean with increasing sample size an effect size will not change. It may increase or decrease depending on what the new data have to say about the overall scientific effect. However, contrary to p-values, **effects sizes won't necessarily increase simply as a function of increasing sample size**. If you are not clear on this difference, you are encouraged to return to Chapter 1 to review these principles or consult Denis (2016, Chapter 3) for a relatively extensive discussion. P-values tell you if you can reject the null and infer to the population, but it is the effect size that gives an indication of mean separation, or difference, or correlation in general (depending on the model you're running). These are all indicators of scientific effect. It's the reason why you're doing the experiment or study in the first place. **You're not doing the study or experiment simply to obtain a small p-value**.

Recall that in ANOVA, η^2 and η_p^2 (where the subscript indicates "partial") were options for assessing effect. We can also compute these in MANOVA. For Wilks' lambda, we can compute η_p^2 as follows:

$$\eta_p^2 = 1 - \Lambda^{1/s},$$

where Λ is the value of Wilks' from the MANOVA, and s is the smaller value of either the number of dependent variables or the degrees of freedom for the independent variable (Schumacker, 2016). For our data, recall $\Lambda = 0.056095$ and degrees of freedom are equal to 2 since we have three levels on the independent variable. Since we also have two dependent variables, s is therefore equal to 2 (i.e. the minimum of 2 dependent variables and 2 degrees of freedom is 2). We compute:

$$\eta_p^2 = 1 - \Lambda^{1/s} = 1 - 0.056095^{1/2} = 1 - 0.23684383 = 0.76$$

How should we interpret the number 0.76? Recall in univariate ANOVA, we would say that 76% of the variance in the dependent variable is accounted for by knowledge of the independent variable. However, we are no longer in a univariate setting. In our analysis, we did not assess dependent variables separately, but rather analyzed them simultaneously as a linear combination. Hence, it is across the linear combination that we are assessing variance explained, not individual dependent variables. There are also some complexities that arise when using η^2 in the MANOVA setting. For details, see Tabachnick and Fidell (2001, p. 339). Effect sizes are also computable for the other multivariate statistics, and may differ from that of Wilk's. See Rencher and Christensen (2012) for a good discussion of effect size in multivariate settings.

9.5   Evaluating Assumptions in MANOVA

The assumptions of MANOVA are similar in many respects to those of ANOVA, but depart in a few important ways. Whereas in ANOVA, the assumption of normality was for that of a single dependent variable across levels of the independent variable, in MANOVA, in addition to each variable considered univariately, the assumption of normality is also on a linear combination of dependent variables. This is what is conveniently called **multivariate normality**.

Regarding the normality assumption of dependent variables considered separately, as was done in ANOVA, one can obtain Q–Q plots or histograms, or, if desired, a formal normality test such as the Shapiro–Wilk test. To verify multivariate normality, the **mvnormtest** package in R will prove useful. To use this test, we need to first **transpose** the data frame (Schumacker, 2016), which means to make rows what are currently columns and columns what are currently rows. We will call the new object iq.train.t to indicate that it has been transposed:

```
> iq.train.t <- t(iq.train)
> iq.train.t
        [,1] [,2] [,3] [,4] [,5] [,6] [,7] [,8] [,9]
quant     5    2    6    9    8    7    9   10   10
verbal    2    1    3    7    9    8    8   10    9
train     1    1    1    2    2    2    3    3    3
```

We are now ready to conduct the multivariate Shapiro test using the mshapiro.test() function:

```
> library(mvnormtest)
> mshapiro.test(iq.train.t)

        Shapiro-Wilk normality test

data:   Z
W = 0.8178, p-value = 0.03254
```

As was true for the univariate Shapiro–Wilk test, the null hypothesis is that the distribution is normal. However, in this case, it is the **multivariate** distribution we are evaluating. The p-value associated with the test is equal to 0.03254, and hence if we set our rejection criterion at 0.05, then this would call for a rejection of the multivariate null. On the other hand, had we set the rejection criteria at 0.01, then we would not reject. We will conclude that the p-value is large enough in this case to deem the assumption satisfied, but because it is quite low, we should keep an eye out for outliers or any other abnormalities in the multivariate residuals.

One further note about normality. Even if you are relatively certain that univariate and bivariate normality are satisfied, it does not guarantee nor imply that multivariate normality is also simultaneously satisfied. What this means is that **univariate and bivariate normality are necessary but not sufficient conditions for multivariate normality**. However, screening univariate plots such as histograms and Q–Q plots as well as bivariate scatterplots and conducting exploratory data analyses in general (i.e. get to know your data well) is a good way, in practice at least, to more or less consider this assumption satisfied. If you detect severe abnormalities, then transformations may be called for. Or, more commonly and pragmatically, do not interpret your obtained p-values too literally. Screening assumptions is not a confirmatory science, and is something you should be doing mostly by visual inspection and exploring as you get to know your data. The idea of "checking off" assumptions is fine, but if you know your data well, you will know how they behave without having to formally do a checklist of things to confirm each assumption is satisfied. Exploratory analyses is always encouraged. **Get to know your data**. **Avoid as much as possible doing "checklist" data analysis**.

9.6 Outliers

To verify there are no univariate or bivariate outliers, histograms and scatterplots can be produced both across and within groups or cells on the independent variable(s). What about multivariate? While outliers can exist in univariate and bivariate space, they can also exist in multivariate space, though they can be extremely difficult to spot graphically. For this reason, computing **Mahalanobis distances**, often just abbreviated to D^2, for each case is generally advised. Mahal distances measure the multivariate distance between each case and the group multivariate mean (often called a **centroid**). We can obtain Mahal distances in R for our IQ data using the **distances** package, via the following for our MANOVA, where specifying `normalize = "mahalanobize"` will give us the required distances. We print for the 9 cases:

```
> library(distances)
> distances(iq.train[1:2], normalize = "mahalanobize")
        1       2       3       4       5       6       7       8       9
1 0.00000 2.0648 0.39413 1.53211 2.69564 2.66667 1.78209 2.42959 2.06476
2 2.06476 0.0000 2.37934 3.03023 2.37934 2.06476 2.75892 3.03023 3.15305
3 0.39413 2.3793 0.00000 1.18967 2.66667 2.69564 1.54891 2.22955 1.78209
4 1.53211 3.0302 1.18967 0.00000 2.22955 2.42959 0.69518 1.33333 0.69518
5 2.69564 2.3793 2.66667 2.22955 0.00000 0.39413 1.54891 1.18967 1.78209
6 2.66667 2.0648 2.69564 2.42959 0.39413 0.00000 1.78209 1.53211 2.06476
7 1.78209 2.7589 1.54891 0.69518 1.54891 1.78209 0.00000 0.69518 0.39413
8 2.42959 3.0302 2.22955 1.33333 1.18967 1.53211 0.69518 0.00000 0.69518
9 2.06476 3.1531 1.78209 0.69518 1.78209 2.06476 0.39413 0.69518 0.00000
```

Inferential tests exist for identifying large Mahal values. However, decisions on what constitutes an outlier here is very subjective (as in all outlier detection), and inferential criteria should definitely not make the decision for you. In practice, one can simply look for values that stand out to learn if any particular case is very deviant from the rest and may be a problem. Plotting the distances using `plot()` may also help in trying to spot relatively large magnitudes (try saving the above object as "d" then plotting using `plot(d)`).

9.7 Homogeneity of Covariance Matrices

Recall that in ANOVA, we had to verify the assumption of homogeneity of variances. That is, we evaluated the tenability of the null hypothesis,

$$H_0 : \sigma_1^2 = \sigma_2^2 = \sigma_3^2 = \sigma_4^2$$

against the alternative hypothesis that somewhere existed a population difference in variances.

In MANOVA, we wish to evaluate a similar hypothesis, but again, the situation is more complex. Why? Because now instead of only variances of each dependent variable, as mentioned earlier, we also have **covariances**. That is, within each level of the independent variable, exists a covariance matrix of dependent variables. The dimension of the covariance matrix will be determined by how many dependent variables we have in the analysis. Hence, instead of verifying only equality of variances, we need to also verify the assumption of **equality of covariances**. Recall that variances and covariances can be packaged nicely into a **covariance matrix**. So, our null hypothesis here for covariances is the following:

$$H_0 : \Sigma_1 = \Sigma_2 = \Sigma_3$$

The above reads that the population covariance matrices between levels on the independent variable are equal, where Σ_1 through Σ_3 represent the covariance matrices for each group.

It doesn't take long to realize that the classic tests for variances that worked so well in the univariate setting will not work in this new and wider multivariate framework. That is, we can't simply apply a Levene's test, since we have covariances to also deal with, and Levene's will only evaluate equality of variances, not covariances. Clearly, we require a different kind of test. The most popular option is to use the **Box-M test**. This test evaluates the assumption of equality of covariance matrices. The details of the test and how it works exactly is beyond the scope of this book, but interested readers may refer to Johnson and Wichern (2007) for a description of the test. Essentially, the test aims to see if there might be an **imbalance** among covariance matrices, and uses determinants and such to evaluate this effect.

In R, we compute the following, where `iq.train[1:2]` identifies the first two variables in our data frame as the dependent variables, and `iq.train[,3]` tells R that the third variable is the train variable:

```
> library(biotools)
> iq.train <- data.frame(iq.train)
> boxM(iq.train[1:2], iq.train[,3])


        Box's M-test for Homogeneity of Covariance Matrices
data:   iq.train[1:2]
Chi-Sq (approx.) = 5.1586, df = 6, p-value = 0.5236
```

The test reports a chi-squared statistic of 5.1586, on 6 degrees of freedom. The associated p-value is equal to 0.5236. Since the obtained p-value for the Box-M test is well larger than some conventional level such as 0.05, we have insufficient evidence to reject the null hypothesis of equality of covariance matrices across groups of the independent variable. That is, we may at least tentatively deem our assumption of equal covariance matrices to be satisfied.

9.7.1 What if the Box-M Test Had Suggested a Violation?

If you obtain a statistically significant result for the Box-M test, then the MANOVA results should be interpreted with a bit of caution in light of this violation. However, and this is important, it is vital to understand that the Box-M test is quite sensitive to sample sizes. Hence, the test should be interpreted with a degree of leniency. The test is also quite sensitive to normality, which is usually violated to some degree. Hence, unless the test suggests a severe violation, under most cases one can proceed with the MANOVA without too much concern.

However, if the violation is rather severe, then you could consider performing a transformation on the dependent variables in an attempt to remedy the situation, such as taking the square root or logarithm of dependent variables and re-analyzing the data. Again however, usually these kinds of modifications are not required, especially given that you are using the MANOVA to look for a **scientific effect**, primarily, rather than using it for **statistical artistry** alone. Since p-values may be unstable somewhat due to violations of assumptions such as this, as always, simply be cautious when interpreting p-values from the MANOVA and do not take the criterion of 0.05 (or whichever one you are using) as "gospel." As always, look to **effect size** to help you discern whether or not there might be something of scientific value occurring in your data.

9.8   Linear Discriminant Function Analysis

In the MANOVA just performed, the objective was to learn if there were mean vector differences of training group on the linear composite of verbal and quant. The fact that we found a statistically significant effect, coupled with a good effect size, suggested that on this linear composite we had evidence of mean vector differences.

There is a technically equivalent way of thinking about the above conclusion, and it is in the following way. Evidence of an effect in MANOVA implies that there are one or more functions generated that is or are "responsible" for the maximal separation that we are observing among the levels of the independent variable. These functions are **weighted linear combinations**, and are known as **linear discriminant functions**. They are the functions, made up of the dependent variables, that are maximally separating groups on the independent variable.

Discriminant function analysis, then, can be considered the **opposite** of MANOVA, in that instead of evaluating mean vector differences, we instead seek to learn more about the linear combination(s) responsible for separation on the independent variable. For this reason, discriminant analysis is often seen as a suitable "follow-up" to MANOVA results. However, especially in the statistical and machine learning literature, it is considered in its own right as a **linear classifier**. Of course, these are (generally) simply different names for similar things. In such literature, discriminant analysis generally falls under the umbrella term of one of many **supervised learning** techniques. The "supervised" element results from having predefined categories in which we wish to classify observations. Later in this book, we survey a few **unsupervised learning** techniques, where the grouping structure is not yet known or at minimum, not specified, or hypothesized. In these techniques, the objective is to "discover" patterns in the data rather than have them hypothesized in advance. Such techniques include principal component analysis and cluster analysis, among others. The idea of "learning" in this sense is merely mathematical in the way of optimizing some function of the data.

Whether one considers discriminant analysis as the opposite to MANOVA will also likely depend on whether a MANOVA was performed beforehand. Though as mentioned, at their technical levels, as we survey them here, the two are virtually indistinguishable at our level of investigation. However, the types of output generated will differ by procedure. What is more, in its simplest case, DISCRIM (discriminant analysis) can be considered the opposite of ANOVA and even a simple *t*-test. **Once you appreciate how many different models essentially generate similar underlying statistical quantities, your interpretation of statistical models will be greatly enhanced, and possibly forever changed.**

> Discriminant analysis is a statistical method that extracts functions of variables that maximally separate groups on another variable, and as such, can be considered the opposite of MANOVA. In MANOVA, the goal was to see if there were mean differences on levels of an independent variable on a composite dependent variable. If there are, then we may wish to learn more about this linear composite that does such a good job at discriminating between levels on the independent variable. This linear composite is the discriminant function. The number of discriminant functions will be equal to either the number of groups on the grouping variable minus 1, or the number of discriminating variables, whichever is smaller.

9.9 Theory of Discriminant Analysis

The theory behind discriminant analysis can be developed in a variety of ways. In 1936, R.A. Fisher, the innovator of the technique, sought out a function that would maximize **group differentiation**. But what does this mean, exactly? We need to define the nature of a linear combination of variables. We will denote a linear combination by ℓ_i:

$$\ell_i = a_1 y_1 + a_2 y_2 + \cdots a_p y_p,$$

where y_1 through y_p are "input" variables or "predictors" in the sense of regression, and a_1 through a_p are weights associated with each respective predictor. In total, there are p weights associated with p variables. Simply put, the job of discriminant analysis is to estimate weights a_1 through a_p that maximize the standardized difference between groups on a response variable. That is, we seek to identify a function, the **discriminant function**, such that when we input values for y_1 through y_p, and couple these with the estimated weights a_1 through a_p, we obtain a function that best discriminates between groups on the response variable.

But what does "best" mean? That is, in what sense are we defining **optimal** or **maximal** discrimination for our function? Recall in least-squares regression we said that the line of best fit, which was really a linear function, was that which minimized the sum of squared errors about the line. That is, the line of best fit was specifically designed to minimize a function of the data, or equivalently, to maximize R^2. In that case, both the minimization of squared errors was technically equivalent to the maximization of R^2. Analogously, in discriminant analysis, the criterion to maximize is between-group differences relative to within-group variation. That is, we seek a function such that when we use it to discriminate among groups, it does so **maximally**. Does that mean the discrimination for any given data set will be good? Not necessarily, not any better than the least-squares line could guarantee a small sum of squared errors for any particular data. However, what is assured on a mathematical level is that the discrimination will be the **best** it can be. That is, it will be **maximized**. Discriminant analysis, as is true of so many other statistical techniques, is a problem of **optimization**.

9.10 Discriminant Analysis in R

We demonstrate a discriminant analysis in R on the exact same data we performed the MANOVA a bit earlier, and while interpreting output unique to discriminant analysis, we also demonstrate how the two techniques have an underlying similarity.

Recall once more our data on verbal, quant, and training group:

```
> iq.train
  quant verbal train
1     5      2     1
2     2      1     1
3     6      3     1
4     9      7     2
5     8      9     2
6     7      8     2
7     9      8     3
8    10     10     3
9    10      9     3
```

Recall that the MANOVA on these data hypothesized quant and verbal as a linear combination of training group. Instead of the MANOVA, let's flip things around and conduct the corresponding discriminant analysis. Our response variable then is now train, and our independent variables are now verbal and quant:

```
> library(MASS)
> lda.fit <- lda(train.f ~ verbal + quant, data=iq.train)
> lda.fit
Call:
lda(train.f ~ verbal + quant, data = iq.train)

Prior probabilities of groups:
     none      some      much
0.3333333 0.3333333 0.3333333
```

```
Group means:
      verbal     quant
none       2 4.333333
some       8 8.000000
much       9 9.666667

Coefficients of linear discriminants:
                LD1            LD2
verbal 0.97946790 -0.5901991
quant  0.02983363  0.8315153

Proportion of trace:
   LD1     LD2
0.9889 0.0111
```

We unpack the above output:

- The `Prior probabilities of groups:` is what we would expect the probability of classification to be per group in the absence of any predictors. In R, this probability is set by default for our data at 0.33 per group, indicating equal priors per group. This prior probability can be adjusted if we had good reason to, but for now, we will leave it at default.
- The `Group means:` represents the mean for each level of `verbal` and each level of `quant`. That is, the mean for verbal = none is 2, the mean for verbal = some is 8, and the mean for verbal = much is 9. Likewise, the mean for quant = none is 4.33, the mean for quant = some is 8.00, and the mean for quant = much is 9.67.
- The `Coefficients of linear discriminants:` are the estimated coefficients for each linear discriminant function. R has produced two discriminant functions here (we will discuss why a bit later). The first discriminant function (LD1) weights `verbal` by `0.97946790` and `quant` by `0.02983363`. The second discriminant function (LD2) weights `verbal` by `-0.5901991` and `quant` by `0.8315153`. The coefficients generated here are the **raw coefficients** for the discriminant functions. We can also generate standardized coefficients. We will see in a moment how we can use these coefficients to manually compute discriminant scores. One can also obtain the coefficients by `lda.fit $scaling`.
- The `Proportion of trace:` values for LD1 of 0.9889 and LD2 of 0.0111 can be considered as measures of **importance** for each function. The relevance of each discriminant function can be computed by contrasting its eigenvalue to the sum of eigenvalues generated by the entire discriminant analysis. R doesn't report eigenvalues, but they are used "behind the scenes" for obtaining the

proportion of trace output. For our data, the first eigenvalue turns out to be 14.35 and the second eigenvalue is equal to 0.16. These two eigenvalues, when summed, make up the trace of the matrix. Therefore, the proportion of the trace accounted for by the first eigenvalue is $14.35/(14.35 + 0.16) = 14.35/14.51 = 0.989$, while the proportion of the trace accounted for by the second eigenvalue is $0.16/14.51 = 0.01$. Clearly, the first discriminant function is much more relevant in discriminating between groups than the second.

We can visualize the **discriminatory power** of these functions by obtaining a plot in R:

```
> plot(lda.fit)
```

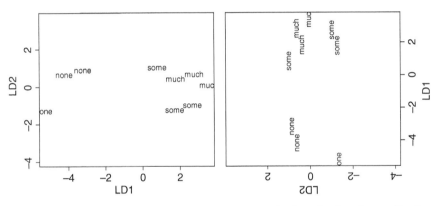

The way to read this plot is by thinking of LDI, which is the first linear discriminant function, as a new variable scaled on the x-axis. That is, LDI is a new **dimension**. Likewise for LD2, it is scaled on the y-axis, and represents a new variable, that is, also a new dimension. Beyond that, simply read the plot as you would an ordinary scatterplot. The points in the plot, with their attached level values (none, some, much) are simply bivariate observations as one would have in any other scatterplot. The difference is that in this plot, they represent joint scores on LD1 and LD2 rather than on traditional variables we would typically have in a scatterplot.

Now, look at LD1 again. Notice that there is a lot of variability of points on this dimension. That is, points are not all lined up at one or two numbers, but rather scattered. We can see from the plot then that LD1, this new dimension, is very good at separating the group "none" from the groups "some" and "much." This then is the graphical representation from what was revealed earlier by our obtained eigenvalues, that the first function is "responsible" for a lot of "separation" among the groups, specifically group "none" from groups "some and much."

Now, consider the second function, LD2. Notice that it is quite poor at discrimination. That is, if you flip the plot (as was done to the right), you can see that LD2 is not doing a great job at discriminating, there is no clear line of division. Hence, the second function, LD2, is quite poor at discrimination, as was suggested by the second eigenvalue.

9.11 Computing Discriminant Scores Manually

In the output, R generated for us the linear discriminant function coefficients for each function:

```
Coefficients of linear discriminants:
              LD1          LD2
verbal  0.97946790  -0.5901991
quant   0.02983363   0.8315153
```

Using these coefficients, R computes discriminant function scores for each case in our data:

```
> predict(lda.fit)
$x
          LD1          LD2
1 -4.3139727   0.61732703
2 -5.3829415  -1.28701971
3 -3.3046712   0.85864322
4  0.7027013   0.99239274
5  2.6318035  -1.01952069
6  1.6225019  -1.26083688
7  1.6821692   0.40219366
8  3.6709386   0.05331078
9  2.6914707   0.64350986
```

To better understand what these numbers actually are, we will compute a few of them manually. For example, how was the first score of -4.3139727 computed? Recall the scores on case 1 in our data, having a 5 for quant and 2 for verbal. Let's demonstrate the computation of the first discriminant score -4.3139727 via the first discriminant function. The intercept (not provided by R, you usually won't need it, we're showing the computation of discriminant scores for demonstration only) is equal to -6.422, and so the computation for the first discriminant score is:

```
Function 1, case 1 = -6.422 + quant(0.030) + verbal(0.979)
                   = -6.422 + 5(0.030) + 2(0.979)
                   = -4.314.
```

Let's compute next the discriminant score for case 1 of the second function:

```
Function 2, case 1 = -2.360 + quant(0.832) + verbal(-0.590)
                   = -2.360 + 5(0.832) + 2(-0.590)
                   = 0.617.
```

We could, if we wanted to, do the same for all the scores in our data. The result would be what R produced in those two columns for LD1 and LD2. These are analogous to **predicted values** in a regression analysis, in that they are generated using the weights from the given discriminant analysis to predict scores on the given discriminant function.

9.12 Predicting Group Membership

Having obtained our discriminant functions, we will now ask R to predict group membership for us, and in this way, we will be able to validate the **goodness** of our functions. We first request predicted probabilities for each group of the dependent variable based on discriminant functions derived:

```
> predict(lda.fit)
$`class`
[1] none none none some some some some much much
Levels: none some much
$posterior
          none             some            much
1 1.000000e+00  1.256554e-08  2.670544e-11
2 1.000000e+00  5.336606e-11  8.295386e-15
3 9.999953e-01  4.693953e-06  3.415652e-08
4 5.788734e-06  6.664762e-01  3.335180e-01
5 1.795062e-11  5.763219e-01  4.236781e-01
6 9.578574e-09  8.232410e-01  1.767590e-01
7 9.938278e-09  5.383859e-01  4.616141e-01
8 1.741694e-14  1.658424e-01  8.341576e-01
9 1.255577e-11  2.540887e-01  7.459113e-01
```

We interpret the above output:

- The $posterior title indicates the **posterior probabilities** associated with prediction into each of the three groups on the response variable. For the first case, we see the posterior probability is near 1.0 (i.e. rounded to 1.0) for the group "none," while it is much less for the other two groups. Recall that 1.256554e-08 is scientific notation, meaning to move back the decimal point by 8 positions.

The number is actually equal to 0.00000001256554. That is, the probability of classification into the group "some" for that first case is exceedingly low. Likewise for the third group "much."

- We can see that when classification is based on the highest probability, cases 1 through 3 are classified into the first group "none," while cases 4 through 7 are classified into the second group "some," and finally cases 8 and 9 are classified into the third group "much."
- It becomes immediately apparent by inspecting the classification results that the discriminant analysis did not result in **perfect prediction**. Otherwise, we would have expected 3 cases per group. We will inspect next where the linear classifier made its errors.

9.13 How Well Did the Discriminant Function Analysis Do?

What R has done so far is generate for us discriminant functions that did their best to maximize group discrimination. However, recall that simply because it maximized group separation does not necessarily mean it did it well. Analogous to least-squares regression, a least-squares line may minimize the sum of squared errors, but for any given data set, it is still a fact that those errors could still be quite large. The least-squares line simply does the best it can, and so does the discriminant function analysis; it does the best it can with the data it has at hand. How good either procedure performs will depend, in part, on the quality of the data at hand.

Hence, having performed the LDA, the next question is to evaluate how well the classifier did its job. We already know based on the above probabilities of group membership that it did not do a perfect job. We'd like a summary of how well it did. We can obtain predicted group membership (Schumacker, 2016) by coding the following in R:

```
> result = predict(lda.fit)$class
> result = cbind(result)
> prior = cbind(train.f)
> out = data.frame(prior, result)
> out
  train.f result
1       1      1
2       1      1
3       1      1
4       2      2
5       2      2
```

6	2	2
7	**3**	**2**
8	3	3
9	3	3

We unpack the above table:

- `train.f` contains the actual membership group, that is, the natural group membership.
- `result` contains the **predicted group membership** based on the previously observed probabilities.
- As we noted, the LDA got the first three cases correct, and also the following three cases, but then misclassified the seventh case. It was supposed to go into the third group ("much") but was classified into the second group instead. This constitutes an **error in classification** or a **misclassified case**. We've manually bolded the misclassified case to emphasize it.

Of course, it would be nice if we could obtain a nice neat summary of how well the LDA performed. We can obtain this easily in R by computing what is known as a **confusion matrix** that depicts the results of the classification:

```
> confusion = table(train.f, result)
> confusion
         result
train.f 1 2 3
   none 3 0 0
   some 0 3 0
   much 0 1 2
```

Let's read off the above table:

- Of those originally in the `none` group on `train.f`, all were correctly classified into the first group on the result, which corresponds also to "none." That is, the LDA got those first 3 cases correct in terms of classification as we discussed previously.
- Of those originally in the `some` group on `train.f`, all were correctly classified into the second group on the result, which corresponds also to "some." That is, the LDA got those 3 cases correct as well in terms of classification.
- However, of those originally in the `much` group on `train.f`, only 2 were classified correctly. One was misclassified into the `some` group on the result (i.e. group 2 on the result). This is the error in classification we noticed earlier.

Overall, the confusion matrix provides a nice, neat way to summarize the results of the classification. What we would have hoped for of course for a perfect LDA is that the diagonal of the matrix would contain all the numbers, and no numbers would appear in the off-diagonal. Hence, perfect classification would have resulted in 3's across the main diagonal. This of course is a very small table, so we have no difficulty in simply looking at the table to notice where the LDA was off. In larger tables, however, it may be more difficult. Fortunately, we can compute in R a table that will reveal to us the proportion of cases correctly classified:

```
> classification = table(prior, result)
> diag(prop.table(confusion))
        1         2         3
0.3333333 0.3333333 0.2222222
```

The above proportions tell us what we already knew, that a proportion of 0.22 was correctly classified into group 3. If we summed the proportions down the main diagonal, and divided by the total number of cases, perfect classification would be indicated by a fraction of 9/9. That is, our hope for the LDA is that 9 cases out of 9 possibilities are correctly classified. How well did our LDA do? Well, it only got 8 out of the 9 possibilities correct. If we sum the proportions down the main diagonal, we should get a number of 8/9 = 0.88. This is exactly what we do get when we use sum(diag(prop.table())) to give us that sum:

```
> sum(diag(prop.table(confusion)))
[1] 0.8888889
```

We see that 0.88 of cases were correctly classified. Of course, we already knew that based on our earlier inspection of the confusion matrix, but these computations are useful here especially in data examples where the confusion matrix may be very large. In our example, we are dealing with a 3×3 matrix. In many problems of classification, the dimension of that matrix may become much larger. It's helpful in R to have computations to summarize the output rather succinctly. We can also produce the relevant proportions in a 3×3 table as follows:

```
> prop.table(confusion)
        result
train.f          1         2         3
   none  0.3333333 0.0000000 0.0000000
   some  0.0000000 0.3333333 0.0000000
   much  0.0000000 0.1111111 0.2222222
```

In addition to seeing the proportions along the main diagonal of 0.33, 0.33, and 0.22, we also see the proportion of 0.11 in the cell in row 3, column 2. That's the 1 out of 9 cases (i.e. 1/9 = 0.11) that was misclassified.

9.14   Visualizing Separation

As always with data, while numbers contain the specifics of our analysis, graphics
and visualization provide powerful depictions of the results. Let's obtain a plot to
visualize the separation achieved by the first discriminant function:

```
> plot(lda.fit, dimen = 1)
> plot(lda.fit, type = "density", dimen = 1)
```

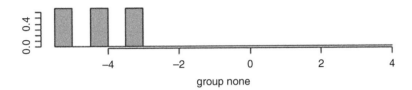

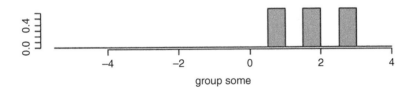

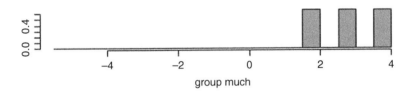

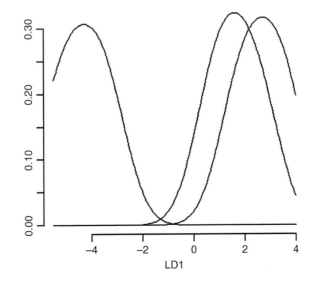

We see once more that much separation is achieved between group "none" and groups "some" and "much." Recall the scores generated by R that we saw earlier (we reproduce them below for convenience):

```
$x
          LD1              LD2
1 -4.3139727   0.61732703
2 -5.3829415  -1.28701971
3 -3.3046712   0.85864322
4  0.7027013   0.99239274
5  2.6318035  -1.01952069
6  1.6225019  -1.26083688
7  1.6821692   0.40219366
8  3.6709386   0.05331078
9  2.6914707   0.64350986
```

As an exercise, see if you can match up the above scores with those in the plots.

9.15 Quadratic Discriminant Analysis

Parallel to MANOVA, one of the assumptions of linear discriminant analysis is that population covariance matrices on each level of the grouping variable are equal. That is, we assume:

$$\Sigma_1 = \Sigma_2 = \Sigma_3 = \cdots \Sigma_p$$

for all populations p. We were able to tentatively evaluate this assumption using the Box-M test in MANOVA, and can likewise apply the test in DISCRIM. If there is a serious violation to the covariance assumption, or there is good **a priori** reason to suspect it is violated, one can perform **quadratic discriminant analysis (QDA)**, in which it is assumed each population has its own covariance matrix, not necessarily equal to one another. Hence, QDA will not require the assumption of equal covariance matrices.

The reason why QDA may be preferable in cases where the covariance assumption is rather severely violated is best summarized by James et al. (2013):

> Why would one prefer LDA to QDA, or vice-versa? The answer lies in the bias-variance trade-off. When there are p predictors, then estimating a covariance matrix requires estimating $p(p + 1)/2$ parameters. QDA estimates a separate covariance matrix for each class, for a total of $Kp(p + 1)/2$ parameters. With 50 predictors this is some multiple of 1275, which is a lot of parameters. By instead assuming that the K classes share a

common covariance matrix, the LDA model becomes linear in x, which means there are Kp linear coefficients to estimate. Consequently, LDA is a much less flexible classifier than QDA, and so has substantially lower variance. This can potentially lead to improved prediction performance. But there is a trade-off: if LDA's assumption that the K classes share a common covariance matrix is badly off, then LDA can suffer from high bias. Roughly speaking, LDA tends to be a better bet than QDA if there are relatively few training observations and so reducing variance is crucial. In contrast, QDA is recommended if the training set is very large, so that the variance of the classifier is not a major concern, or if the assumption of a common covariance matrix for the K classes is clearly untenable. (pp. 149–150)

We will now perform a QDA on our previous data on which we performed LDA. The function in R used to fit QDA is qda ():

```
> library(MASS)
> qda.fit <- qda(train.f ~ verbal + quant, data = iq.train)
> qda.fit

Call:
qda(train.f ~ verbal + quant, data = iq.train)

Prior probabilities of groups:
      none       some       much
0.3333333  0.3333333  0.3333333

Group means:
      verbal      quant
none       2   4.333333
some       8   8.000000
much       9   9.666667
```

The above is the same output we obtained when performing LDA. Where the output will differ is when we use predict (qda.fit), which will yield different probabilities of membership on the grouping variable. Just as we did for LDA (Schumacker, 2016), let's see how our QDA did in terms of predictions into the relevant classes:

```
> result = predict(qda.fit)$class
> result = cbind(result)
> prior = cbind(train.f)
> out = data.frame(prior, result)
> out
```

```
    train.f result
1        1       1
2        1       1
3        1       1
4        2       2
5        2       2
6        2       2
7        3       3
8        3       3
9        3       3

> confusion.qda = table(train.f, result)
> confusion.qda
         result
train.f 1 2 3
   none 3 0 0
   some 0 3 0
   much 0 0 3
```

We see then that for this data, the classification is actually better than for LDA. That is, QDA for this small sample has correctly classified all cases.

9.16 Regularized Discriminant Analysis

There are other varieties of discriminant analysis that like QDA, relax some of the assumptions of traditional LDA. **Regularized discriminant analysis** can be considered a kind of compromise between linear discriminant analysis and the QDA we just featured. Recall that QDA allows for separate covariances when the assumption of equality could not be made as required in traditional LDA. Regularized discriminant analysis shrinks these separate covariance matrices toward a common covariance matrix, more similar to LDA (Hastie et al. 2009). Though not demonstrated here, the package **klaR** can be used to fit such models and should be consulted for more details.

Exercises

1 Describe how multivariate analysis of variance differs from univariate ANOVA. What makes it "multivariate?"

2 What should dominate the rationale for performing a MANOVA? Should it be substantively driven or statistical? Discuss.

3 Describe what is meant by a linear combination, and how such figures in both MANOVA and discriminant analysis.

4 Discuss the relationship between Pillai's Trace and Roy's Largest Root. How are they similar? Different?

5 Suppose you perform a MANOVA, then follow up the MANOVA with several univariate ANOVAs. What are some issues that may present themselves with this approach?

6 Consider the following small data set on variables y, x1 and x2:

y	x1	x2
0	4	2
0	3	1
0	3	2
0	2	2
0	2	5
1	8	3
1	7	4
1	5	5
1	3	4
1	3	2

Perform a MANOVA where x1 and x2 are response variables, and y is an independent variable having two levels. Provide a brief summary of findings.

7 Compute an effect size for the analysis in #6. How much variance is accounted for by the grouping variable?

8 Evaluate the assumption of homogeneity of covariance matrices for the analysis in #6 using the Box-M test.

9 Recall the challenger data of the previous chapter. Perform a discriminant analysis where oring is the binary response variable, and temp is the predictor. Provide a brief summary of findings.

10 Generate a confusion matrix for the analysis in #9. What proportion of cases were correctly classified? How many were misclassified?

10

Principal Component Analysis

LEARNING OBJECTIVES

- Understand the nature of principal component analysis, and perspectives on how it may be useful.
- Understand the differences between principal component analysis and exploratory factor analysis.
- How to run and interpret a component analysis in R using `prcomp()` and `princomp()` functions.
- How to plot correlation matrices among variables to assess magnitudes of correlation.
- How to make potential substantive sense of components derived from PCA.

Principal component analysis, or "PCA" for short, is a **data reduction** technique used to account for variance in a set of p variables into fewer than p dimensions. Traditional PCA assumes your variables are **continuous** in nature. As an example, suppose a researcher has 100 variables in his or her data set. In these variables is a certain amount of variance. That is, the entire set of variables contains a particular amount of variability. If we wanted to know the **total variance** of the variables, we could sum them up and obtain a measure of the total variance. What PCA seeks to do is take this original set of variables in p dimensions, and performs a transformation on these variables such that the total amount of original variance is preserved. For example, if the 100 variables accounted for a total variance of 50, the PCA transformation will not change this total amount of variance. What it will do is attempt to explain this variance in as few new "components" as possible, where the first few components extracted will account for most of the variance in the original set of variables. At least that is the hope, if the technique is successful on a substantive level. The goal is to have most of the variance load on the first few components and hence "reduce the data" down to these few components. Instead of needing all of the variables, the researcher can instead focus on the first few components that account for most of the variance in the variables.

As an example of what PCA accomplishes, consider the following plot adapted from Denis (2016):

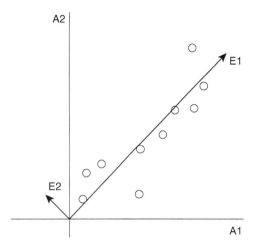

In the plot, the original axes are A1 and A2. The "new" axes after the transformation are E1 and E2. Though there are several ways of interpreting how PCA works, one common way is viewing the sample components as generated by rotating the coordinate axes so that they pass through the scatter in the plot of maximum variance (Johnson and Wichern, 2007). Hence, we can see that what PCA has done is performed a **rotation** such that the new axes optimize some function of the data. What function is maximized? As we will see, the rotation occurs such that the **first component accounts for as much variance in the original variables as possible**, while the **second component accounts for as much of the variance in the original variables as possible, but subject to the fact that it is orthogonal to the first component**. Notice that in the figure, the axes are perpendicular to one another, which is another way of saying that they are at 90° angles. This is the condition of orthogonality.

10.1   Principal Component Analysis Versus Factor Analysis

Before we begin, we need to issue a caveat and warning. In the next chapter, we will be studying something called **exploratory factor analysis (EFA)**. For a variety of reasons, PCA and EFA are often relegated to being the same or at least strikingly similar techniques. You may hear in your communications with some such things as "Oh, whether you do a component analysis or a factor analysis you'll get the same results." However, it is imperative, from the outset, that you do not equate component analysis with factor analysis, as they are not the same procedure, and there are very important and key distinguishing features between them. We cannot discuss these important differences yet, because we haven't surveyed either technique sufficiently, but we will a bit later after we have some of the basics under our belts. Until then, however, take it on trust that the two techniques are sufficiently different that they merit their own separate chapters.

Principal component analysis (PCA) and exploratory factor analysis (EFA) are not equivalent procedures and should not be treated as such. There are key differences between them and the kinds of substantive (scientific) conclusions that can be drawn from each.

10.2   A Very Simple Example of PCA

PCA is usually applied to data sets that contain many variables, sometimes hundreds, since it is for these data sets especially that we wish to reduce dimensionality. For example, if our data set has 1000 variables, then PCA might be useful to reduce the dimensionality down to 2 or 3 dimensions (if the procedure were successful). In this way, data like this makes PCA a potentially suitable analytic approach because of the numerous variables or dimensions we start out with. Other times we wish to conduct a PCA on data sets with a smaller number of variables, for instance, 5–10, to see if we can likewise reduce or uncover the dimensionality of the data. Regardless of how many variables you may be starting out with, PCA may be a suitable technique for your data.

To understand PCA from first principles, however, it is very useful and pedagogical to start off with a very easy example featuring only two variables. Why begin with an example with only two variables? We do so because it allows one to appreciate a bit more what PCA actually **does** instead of getting lost in a more complex example where we may fail to appreciate its underlying mechanics. With techniques such as PCA and EFA, it is far too easy to get immersed in one's substantive theory and coincidentally conclude exactly what you set out to find! That's why it's important to study PCA on its own merits first and foremost, so one can appreciate what the technique can versus cannot accomplish.

What is more, our first example of a PCA is actually one featured by a pioneer of component analysis, Karl Pearson, infamous statistician who first introduced the technique in a paper written in 1901. Once we can appreciate how PCA works on such a small sample, we will be in good shape to consider even the most complex of data sets on which component analysis may be well-suited.

10.2.1 Pearson's 1901 Data

As our first easy example of PCA then, consider the data given by Karl Pearson in 1901 on two variables, x and y. We will designate these variables as simply x and y without giving them substantive or research names, so we can observe quite simply what PCA does with these variables:

x	y
0.00	5.90
0.90	5.40
1.80	4.40
2.60	4.60
3.30	3.50
4.40	3.70
5.20	2.80
6.10	2.80
6.50	2.40
7.40	1.50

We first create a data frame in R for these two variables:

```
> x <- c(0, 0.9, 1.80, 2.60, 3.30, 4.40, 5.20, 6.10, 6.50, 7.40)
> y <- c(5.90, 5.40, 4.40, 4.60, 3.50, 3.70, 2.80, 2.80, 2.40, 1.50)
> pca.data <- data.frame(x, y)
> pca.data
     x   y
1  0.0 5.9
2  0.9 5.4
3  1.8 4.4
4  2.6 4.6
5  3.3 3.5
6  4.4 3.7
7  5.2 2.8
8  6.1 2.8
9  6.5 2.4
10 7.4 1.5
```

Let's obtain a plot of these data:

```
> plot(x, y)
```

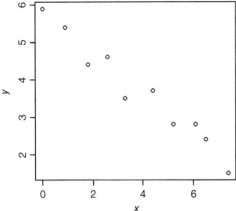

These data are, of course, in two dimensions, x and y. The question PCA asks is:

Can we transform these variables on two dimensions into two new variables also on two dimensions, but such that most of the variance in the variables is accounted for by the first dimension only?

That is, as mentioned at the outset, PCA will seek to perform a transformation on the variables, with the hope that the first dimension will account for most of the original total variance. Be careful to note here that nowhere have we yet reduced the number of dimensions. In the transformation, **PCA will generally generate as many dimensions as there are original variables** in the data set. However, the hope is that most of the variance in the original variables will be accounted for by the first dimension in this case. Had we 100 original dimensions, we would have likewise wanted to transform these into 100 new dimensions, with the hope that most of the variance in the original dimensions (variables) could be accounted for by the first few components. This is the nature of PCA – it can be interpreted as nothing more than a transformation of the original axes onto new dimensions. The characteristics of these new dimensions (components) and how many of these we may choose to retain is another story (a story we will unpack shortly).

To perform the PCA, we first need something to represent how the variables x and y **covary** together. For this, we will build the **covariance matrix** of x and y. In R, we will call this object by the name of A:

```
> A <- cov(pca.data)
> A
         x          y
x  6.266222  -3.381111
y -3.381111   1.913333
```

Recall the nature of a **covariance matrix**. It contains variances of the variables along the main diagonal going from top left to bottom right. For our data, the variance of x is equal to 6.26, and the variance of y is equal to 1.91. The covariances between x and y are given in the off-diagonal of the matrix. We can see in the matrix that the covariance between x and y is equal to -3.38. Notice these numbers are the same in the matrix, since there is only a single covariance between x and y. That is, the number in the bottom left of the matrix is the same as the number in the upper right. The covariance matrix is an example of a **symmetric matrix**, meaning that the upper triangular part of it is the same as the lower triangular part. The reason why we have built a covariance matrix of x and y is because in a moment we will ask R to perform a component analysis directly on this matrix.

For demonstration, and to confirm that we created our covariance matrix correctly (as well that R computed it correctly), we compute the variances and covariance of x and y manually in R. First, we confirm that the variances have been computed correctly:

```
> var(x)
[1] 6.266222
> var(y)
[1] 1.913333
```

We see that the variances computed match those in the covariance matrix. We could have also used the `diag()` function on the matrix to obtain the values along the main diagonal, which of course correspond to the above variances:

```
> diag(A)
       x        y
6.266222 1.913333
```

Next, let's compute the covariance:

```
> cov(x,y)
[1] -3.381111
```

We confirm that the covariance of -3.38 matches that computed in the covariance matrix.

10.2.2 Assumptions of PCA

Because PCA does not invoke a statistical model as do most other methods in this book, used descriptively at least, the assumptions for PCA are quite limited. These are best not regarded as true "assumptions" at all such as we would have in ANOVA or regression, for example, where we are estimating parameters. As discussed, PCA at its core is simply a transformation of data onto new dimensions. No assumptions are needed to perform such a transformation. However, certain conditions should be in place before such a transformation can be performed. For one, the covariance or correlation matrix should contain sufficient covariance or correlation to make the PCA worthwhile. For instance, if the correlation matrix were an **identity matrix**, having values of 1 along the main diagonal and zeroes everywhere else, then performing PCA would make little sense.

We can verify that the correlation matrix is not an identity matrix by conducting what is known as **Bartlett's test of sphericity**. To run the test, we can use the **rela** package, and will employ the function paf(). Along with this test will also be reported the **Kaiser–Meyer–Olkin (KMO) measure of sampling adequacy**. The KMO is a measure of the likely utility of conducting a factor analysis in the sense of providing an estimate of the proportion of variance in variables that might be due to underlying factors. It is less relevant for components analysis, though we nonetheless demonstrate it here in conjunction with Bartlett's. Values higher than 0.70 (or so) are preferred, though even if one obtains a much smaller value (we obtain 0.5 below), this does not imply somehow that the analysis should absolutely not be conducted. We adapt our code again from Schumacker (2016):

```
> library(rela)
> paf.pca = paf(pca.data,eigcrit=1, convcrit=.001)
[1] "Your dataset is not a numeric object."
```

You may receive this error message when attempting to run the paf() function. To make things work here, simply convert the data frame pca.data to a matrix in this case, then try the paf() function again on this new matrix:

```
> pca.data.matrix <- as.matrix(pca.data)
> paf.pca = paf(pca.data.matrix,eigcrit=1, convcrit=.001)
> paf.pca

$KMO
[1] 0.5

$Bartlett
[1] 23.013
```

We have printed only partial output from the paf() function, revealing the KMO and Bartlett values. KMO is a bit low at 0.5, but again, there is nothing wrong with proceeding with the PCA/factor analysis anyway. Bartlett's is equal to 23.013. To get a significance test on this, we compute (Schumacker, 2016):

```
> cortest.bartlett(pca.data.matrix, n = 10)
$`chisq`
[1] 23.013

$p.value
[1] 1.6092e-06

$df
[1] 1
```

The null hypothesis for Bartlett's is that the correlation matrix is an **identity matrix**. The alternative hypothesis is that it is not. Since we have obtained a very small *p*-value, we can safely reject the null and conclude the alternative that the matrix is not an identity matrix. That is, we have evidence to suggest we have sufficient correlation in the matrix to proceed. Again, however, it needs to be emphasized that these measures are simply guidelines and nothing more. There is nothing inherently wrong with performing a component/factor analysis on data that do not meet the above guidelines. In a matrix that resembles an identity matrix, for instance, the resulting analysis will likely not be successful, but that does not mean the procedure cannot still be run to confirm this. The above tests do not by themselves restrict one from running the procedure.

10.2.3   Running the PCA

Having produced our covariance matrix, we are now ready to run the PCA in R. We will first use R's princomp() function to obtain PCA results, then later demonstrate an alternative function in R to obtain the PCA:

```
> pca <- princomp(covmat = A)
> summary(pca)

Importance of components:
                          Comp.1       Comp.2
Standard deviation     2.8479511 0.262164656
Proportion of Variance 0.9915973 0.008402695
Cumulative Proportion  0.9915973 1.000000000
```

We summarize the above output:

- The standard deviation of the first component is equal to 2.8479511. To get the variance, we square the standard deviation and obtain 8.11. Directly below this number, the Proportion of Variance accounted for by this first component is 0.99, which for the first component, is also equal to the cumulative proportion as well.

- The standard deviation of the second component is equal to 0.262164656. To get the variance, we again square the standard deviation and obtain 0.0687. Directly below this number is the proportion of variance accounted for by this second component, equal to 0.008. The cumulative proportion across both components is 1.0.

How were the above numbers computed? The total variance in the variables is equal to their sum, so when we sum the variance across the two variables, we have: $6.266222 + 1.913333 = 8.18$. The first component accounts for a certain proportion of this total variance. Since its variance is 8.11, it accounts for a proportion of $8.11/8.18 = 0.99$. Likewise for the second component, it accounts for a proportion of $0.0687/8.18 = 0.008$. Hence, notice what the PCA has done – it has taken the two original variables or dimensions, and transformed them into two new dimensions on which the variance of the original variables has been **redistributed** across the new dimensions. **Clearly, the PCA has redistributed most of the variance in the original variables to the first component, such that instead of having to interpret both original variables, we can accomplish pretty much just as much by interpreting only the first component without loss of much "information."** This is the essence and purpose behind PCA. The total variance has been preserved, but now most of it is accounted for by the first derived component. This is why the first component is called, appropriately, the "principal" component, in that it accounts for most of the variance in the original variables, while, as we will soon discuss, being **orthogonal** with remaining components.

Of note, we could have also gotten the PCA by specifying the variables in our analysis directly rather than coding the covariance matrix, using `prcomp()`:

```
> fit.pca <- prcomp( ~ x + y)
> fit.pca
Standard deviations (1, .., p=2):
[1] 2.8479511 0.2621647

Rotation (n x k) = (2 x 2):
          PC1          PC2
x   0.8778562  -0.4789243
y  -0.4789243  -0.8778562
```

Notice in the above code that the function `prcomp()` is acting not on a covariance matrix in the call statement, but rather directly on the variables x and y. Regardless of the approach, the result is the same. However, as one can imagine, especially for data sets with a very large number of variables, listing each variable as we did above by ~ x + y would be very time consuming for anything other than a minimal number of original variables, and so constructing the relevant covariance matrix is usually a preferable strategy.

10.2.4 Loadings in PCA

Now we said that PCA performed a transformation on the original variables such that most of the variance in the original variables is now attributed to the first component, while a much smaller amount of variance in the original variables is attributed to the second component. The next question is,

What is the nature of these new components?

That is, we wish to know their "make-up." What is the first component actually composed of? What is the second component composed of? It would make sense that the new components are actually made of, in some sense, the original variables. When you think about it, there really isn't any other option, since the new components can only be composed of variables originally submitted to the component analysis. This fact in itself is another example of why **the most important part of any statistical analysis are the variables you subject to that analysis**. So, when we are asking the question about the make-up of a component, that component can only be made up of the variables you derived it from. Now, this can seem like a no-brainer, but it is a no-brainer that is often overlooked and underappreciated. If you add or subtract one or more variables from the original set, it stands that the component structure will change. Hence, when you draw a conclusion from your component analysis that you have "found" a component in your data, that "finding" is only as good as the variables subjected to the component analysis, and not independent of it. Obvious, right? But very important to remember. Again, component analysis is not "discovering" components, it is merely issuing a transformation of the original information subjected to it. It, as is true of any statistical technique, is only as good as the data on which it is applied.

The most important element of virtually any statistical analysis, including principal component, is the data subjected to that analysis. If you add one or two or more variables, the solution will likely change, and so will your substantive conclusions. The selection of variables and quality of data, which includes measurement and psychometric properties, is always paramount to any statistical analysis. Statistical analyses cannot change bad data into good data.

The new components are what are referred to as **linear combinations** of the original variables. Though the idea of these linear combinations is similar in spirit to that studied in MANOVA and discriminant analysis, the nature of these linear combinations is different, in that they have different properties than those featured in discriminant analysis. Hence, be sure not to equate linear combinations in one procedure with linear combinations in the other since they are not constructed the same way. The linear combinations in discriminant analysis are constructed such that they maximize group differences on the grouping variable. The linear combinations in PCA are constructed such that they maximize variance on the first few components extracted. The degree to which the original variables "contribute" to each of the respective components is measured by the extent to which each original variable "loads" onto that component. This will make much more sense in a moment. For now, let's ask R to generate loadings of the PCA we performed earlier:

```
> pca$loadings
```

```
Loadings:
   Comp.1  Comp.2
x   0.878   0.479
y  -0.479   0.878
```

```
                 Comp.1  Comp.2
SS loadings        1.0     1.0
Proportion Var     0.5     0.5
Cumulative Var     0.5     1.0
```

We summarize what the above tells us:

- Component 1 has a loading for x equal to 0.878, and a loading of -0.479 for y. That is, the first component is the following linear combination:

  Component 1 $= 0.878(x) - 0.479(y)$

- Component 2 has a loading for x equal to 0.479, and a loading of 0.878 for y. That is, the second component is the following linear combination:

  Component 2 $= 0.479(x) + 0.878(y)$

- The SS loadings for each component refer to the sums of squared weights for each column of loadings. One constraint of PCA is that the sum of these squared loadings for each component sum to 1.0. This is a mathematical constraint set by design in the procedure so that the variances of components cannot grow without bound. That is, for components 1 and 2, respectively, we can verify in R:

```
> (0.878)^2 + (-0.479)^2
[1] 1.000325
> (0.479)^2 + (0.878)^2
[1] 1.000325
```

10.3 What Are the Loadings in PCA?

You might be asking at this point how the above loadings are obtained. We know they make up elements of each component, but we have not yet discussed how they are actually derived. To understand how they are derived, we need to return to the concepts of eigenvalues and eigenvectors first discussed in Chapter 2. Though technically speaking there are differences between how `prcomp()` and `princomp()` obtain eigenvalues and eigenvectors (`prcomp()` uses a technique called **singular value decomposition**, while `princomp()` uses eigen decomposition), a common theme is the extraction of eigenvalues and eigenvectors, which we briefly review with reference to eigen decomposition.

Recall that for every square matrix $\mathbf{A}$, it is a mathematical fact that a scalar λ and vector $\mathbf{x}$ can be obtained so that the following equality holds true:

$$\mathbf{A}\mathbf{x} = \lambda\mathbf{x}$$

What this meant is that as we perform a transformation on the vector $\mathbf{x}$ via the matrix $\mathbf{A}$, this equated to transforming the vector $\mathbf{x}$ by multiplying it by a "special scalar" λ. This scalar we called an **eigenvalue** and $\mathbf{x}$ was referred to as an **eigenvector**.

In this brief recollection of eigen analysis, we have effectively summarized the technical underworkings of PCA. When using eigen decomposition, PCA features an **eigen analysis**, not on any arbitrary matrix, but rather on the covariance matrix containing the variables we subjected to the PCA. There is a wealth more to learn about eigen analysis, and the interested reader is encouraged to consult any book on linear algebra for more details, or a comprehensive book that discusses the theory of multivariate analysis more extensively, such as Johnson and Wichern (2007) or Rencher and Christensen (2012). **To the mathematician,**

many multivariate statistical methods reduce, on a structural level, to eigen analysis.

In R, we can demonstrate the computation of the relevant eigenvalues and eigenvectors that we obtained in the above PCA by requesting an eigen decomposition directly:

```
> eigen(A)
eigen() decomposition
$`values`
[1] 8.11082525 0.06873031

$vectors
            [,1]         [,2]
[1,] -0.8778562 -0.4789243
[2,]  0.4789243 -0.8778562
```

We can see that the eigenvalues generated of 8.11 and 0.068 match those obtained from our PCA. We can also see that other than a change in sign, the magnitude of the loadings are the same. Again, this is as far as we take the topic of eigenvalues and eigenvectors in PCA. The interested reader is referred to the aforementioned sources for more details. The key point for now is to appreciate how universal eigen analysis is across both mathematics and the sciences, and is the workhorse of **dimensionality reduction**.

10.4 Properties of Principal Components

In our example above, we derived the components, but we only touched on a few of the properties specific to them. We now summarize and extend on that discussion here. Components obtained in a PCA obey the following properties:

1) **The sum of the obtained eigenvalues of the derived components from the covariance matrix matches the sum of the original variances of the raw variables.** For our data, recall the sum of the eigenvalues was equal to 8.18. This number is the same that we obtain when summing the original variances, those of 6.27 and 1.91 = 8.18. Why is this property important? It is vital to understanding what PCA does because it reveals that PCA does not change the amount of "information" in the original variables in terms of how much variance they account for, it merely transforms this information (variance) onto new dimensions. That is, **principal components preserves**

the **original total variance in the variables**. It simply redistributes this variance across new components, that is, across new dimensions.

2) **Successive components (which recall, are linear combinations of the original variables) are orthogonal**. When we obtain principal components, the first component obtained is that which explains the maximum amount of variance in the original variables as possible. This is the first principal component. The second component obtained explains the maximum amount of variance in the original variables, however, subject to the constraint that it be orthogonal to the first component. In this way, extracted components do no overlap.

3) **The sum of squared loadings for each component equals 1.0.** That is, the variance of each component is maximized subject to the constraint that the length of each component be equal to 1.0. Why is this constraint relevant? It is important because if we did not subject the components to some constraint such as this, then in theory, the **variance of the given component could grow without bound**. That is, the variance of the component could get larger and larger if we did not have some way of regulating it. Constraining the sum of squared loadings to equal 1.0 is one way of making sure the variance of each component is "kept in check" in some sense. One caveat: in some software programs such as SPSS, the sum of squared loadings in its program for PCA is typically not equal to 1.0, but rather will equal the corresponding eigenvalue for the given component. There are reasons for this that are beyond the scope of this book (they relate to how SPSS subsumes PCA under the technique of factor analysis rather than its own independent method) but it's the principle that is worth remembering for now, and that is putting some kind of constraint on the length of the eigenvector.

10.5 Component Scores

Having generated two components, we now wish to obtain **component scores** on each one. That is, for any given case in our data, we wish to know the component score for that given individual on each component. We can obtain component scores by computing them manually on each component as a linear combination of the observed variables x and y. For the first component then, we compute:

```
> comp.1 = 0.8778562*(x-3.82) -0.4789243*(y-3.7)
> comp.1
```

```
[1]  -4.4070441 -3.3775114 -2.1085165 -1.5020164 -0.3607004
0.5091566
[7] 1.6424734 2.4325440 2.9752562 4.1963587
```

Note carefully how the above was computed: we simply weighted x and y by the corresponding weights for the eigenvector designated for that component. We subtracted the mean of each variable in doing so, because by default, prcomp() typically computes components by first centering x and y to have mean zero (James et al. 2013).

Likewise, we can compute scores for the second component:

```
> comp.2 = 0.4789243*(x-3.82) +0.8778562*(y-3.7)
> comp.2

[1]   0.10179281   0.09389658 -0.35292775   0.20578293
-0.42461188   0.27777609
[7] -0.12915505   0.30187682   0.14230406 -0.21673465
```

Now if we have computed these components correctly, they should have variances equal to the eigenvalues we extracted from the PCA. We can easily check this by computing the variances of each component:

```
> var(comp.1)
[1] 8.110825
> var(comp.2)
[1] 0.06873031
```

Notice that the above variances for each component match the respective eigenvalues from the PCA.

10.6 How Many Components to Keep?

10.6.1 The Scree Plot as an Aid to Component Retention

The big question in PCA from a **scientific** rather that statistical perspective, is usually how many components to keep. This is probably the most substantively difficult part of the entire PCA, because **you cannot rely on the statistical method to tell you how many you should keep**. Instead, the decision is often made rather **subjectively**, but with the aid of some statistical tools. One such tool is the **scree plot**, a graphical device that plots the eigenvalues (variances) for each component. In a scree plot, one looks for a "bend" in the plot to help determine the number of components to retain. We can obtain a scree plot for our data as follows:

```
> screeplot(pca, type="lines", col=3)
```

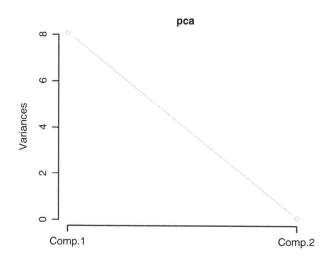

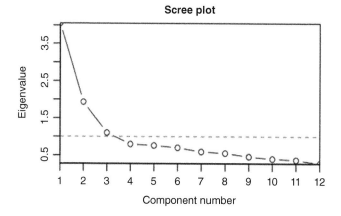

Now of course, having only two components possible for the current analysis, the first plot does not truly reveal any real "insight" into how many components to keep. There is no bend in the plot when there are only two components. Again, however, with such a simple example, this makes it possible to see what the scree plot actually accomplishes, which is simply graphing the eigenvalues for each component. By itself, it doesn't really tell us much more than that. When we conduct an analysis with more than two components a bit later, the plot will look a bit more complex. The second plot, for instance, is a scree plot from a more complex analysis. Arguably, the bend occurs roughly at an eigenvalue of approximately 2 or 3, which according to **Kaiser's rule**, would recommend to keep components with eigenvalues greater than 1.0 when analyzing a correlation matrix. The logic behind Kaiser's rule, that of keeping components with eigenvalues greater than 1.0 in a correlation matrix, is that any component having an eigenvalue more than 1.0 accounts for more than the average total variance (since each variable contributes a unit variance). In a correlation matrix, the average variance is equal to 1.0. In a covariance matrix, this will not necessarily be the case, but the rule can still be applied, that of retaining components that exhibit greater that average variance (whatever that number turns out to be, it will likely not be 1.0 as was true for the correlation matrix). Kaiser's rule is a guideline at best, and in no way should be used as "hard evidence" regarding how many components to retain. If as a researcher you're retaining components based only on Kaiser's rule, your thought process regarding component analysis is incomplete.

10.7 Principal Components of USA Arrests Data

We demonstrate another PCA, this time on the **USA Arrests** data, which is a data frame available in the base package in R. The variables in this data sets are `murder`, `assault`, `urbanpop`, and `rape`. The cases in the analysis are the states. Following James et al. (2013), we detail the PCA for these data, and refer you to their analysis (pp. 373–385) to highlight one important element of principal components, that **they are unique only up to a change in sign**. Hence, the positives and negatives on component loadings could change depending on what programs are used to obtain the components (as we saw earlier), but this is not indicative of results being "wrong" or contradictory. This is what we mean by saying **the components are unique, but only up to a change in sign**.

Before conducting the analysis, let's get a look at our data:

```
> head(USArrests)
         Murder Assault UrbanPop Rape
Alabama    13.2     236       58 21.2
Alaska     10.0     263       48 44.5
Arizona     8.1     294       80 31.0
Arkansas    8.8     190       50 19.5
California  9.0     276       91 40.6
Colorado    7.9     204       78 38.7
```

Let's obtain a matrix of correlations for these data:

```
> cor(USArrests)
              Murder    Assault   UrbanPop       Rape
Murder    1.00000000 0.8018733 0.06957262 0.5635788
Assault   0.80187331 1.0000000 0.25887170 0.6652412
UrbanPop  0.06957262 0.2588717 1.00000000 0.4113412
Rape      0.56357883 0.6652412 0.41134124 1.0000000
```

We can also get the correlations individually if we wanted, by specifying the data frame, then requesting which variables we want to correlate:

```
> cor(USArrests$Murder, USArrests$Assault)
[1] 0.8018733
```

Recall that the function of $ here is to select the given variables from the data frame that we wish to correlate.

Inspecting a correlation matrix of numbers is one thing, but an even more powerful way to interpret data is via **visualization**. Below we produce the corresponding scatterplots of these correlations, as well as a more elaborate plot containing both bivariate and univariate information (down the main diagonal):

```
> pairs(USArrests)
> library(GGally)
> ggpairs(USArrests)
```

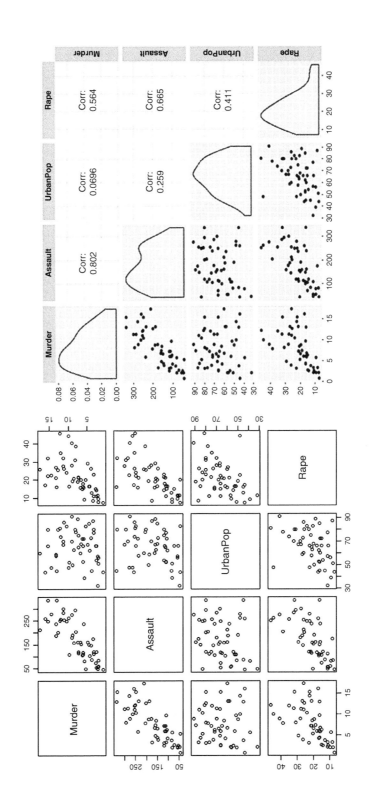

We use the `prcomp()` function to get the PCA:

```
> usa.comp <- prcomp(USArrests, scale=TRUE)
> usa.comp
Standard deviations (1, .., p=4):
[1] 1.5748783 0.9948694 0.5971291 0.4164494

Rotation (n x k) = (4 x 4):
                PC1         PC2         PC3         PC4
Murder   -0.5358995  0.4181809 -0.3412327  0.64922780
Assault  -0.5831836  0.1879856 -0.2681484 -0.74340748
UrbanPop -0.2781909 -0.8728062 -0.3780158  0.13387773
Rape     -0.5434321 -0.1673186  0.8177779  0.08902432
```

- By specifying `scale=TRUE`, we've requested that the analysis be done on variables with standard deviation 1. That is, the variables are scaled to have unit variance, and by default `prcomp()` centers the variables to have mean zero (James et al. 2013). The analysis is being performed on the **correlation** matrix rather than the **covariance** matrix.
- Four variables were subjected to the analysis, and so four components were generated, each having standard deviations equal to 1.57, 0.99, 0.59, and 0.41.
- The loadings for each component are given next. It would appear at first glance that `murder`, `assault`, and `rape` load moderately well on PC1, while `urbanpop` does not, but does load strongly on PC2. `Rape` loads quite high on PC3, and `murder` and `assault` load fairly high on PC4.

Next, we obtain a scree plot of the component analysis (Kassambara, 2017).

```
> library(FactoMineR)
> library(factoextra)
> scree <- fviz_eig(usa.comp)
> scree
```

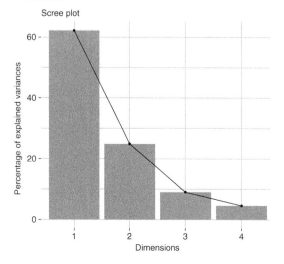

We can see from the plot that the first component extracted indeed accounts for much of the variance in the original variables.

10.8   Unstandardized Versus Standardized Solutions

PCA can be performed on either the original **unstandardized** data, or on data **standardized** to have mean equal to zero and standard deviation equal to 1. This transformation is equivalent to analyzing the **correlation matrix** (standardized data) rather than the **covariance matrix** (Izenman, 2008).

The choice of whether or not to standardize is not a simple one, and the decision to standardize should not be made without careful consideration. If the original variables are not measured in the same units, and have wildly different variances and are hence not considered "commensurate," then this could have a rather dramatic effect on the PCA solution. The loadings for the variables with the greatest variance may be exaggerated in a sense since they contribute the most original variance. However, as noted by Rencher and Christensen (2012):

> Generally, extracting components from **S** [covariance matrix] rather than **R** [correlation matrix] remains closer to the spirit and intent of principal component analysis, especially if the components are to be used in further computations ... However, in some cases, the principal components will be more interpretable if **R** is used. (pp. 419–420).

As noted by Jolliffe (2002), standardization does have the potential drawback of making statistical inference with regard to components more difficult, though since PCA is typically used as a descriptive rather than inferential tool, this should not by itself impede the decision to standardize. Other authors on the subject are even more adamant about virtually always standardizing:

> Because it is undesirable for the principal components obtained to depend on an arbitrary choice of scaling, we typically scale each variable to have standard deviation one before we perform PCA. (James et al. 2013, p. 381)

As a demonstration of how a change in scale can influence the results of a PCA, consider a comparison of component analyses on the USArrests data analyzed by covariance versus correlation matrix. To analyze on the covariance rather than the correlation matrix (as we did earlier), we specify scale = FALSE:

```
> usa.comp.cov <- prcomp(USArrests, scale=FALSE)
> usa.comp.cov
```

```
Standard deviations (1, .., p=4):
[1] 83.732400 14.212402   6.489426   2.482790

Rotation (n x k) = (4 x 4):
                      PC1            PC2            PC3            PC4
Murder     0.04170432 -0.04482166   0.07989066 -0.99492173
Assault    0.99522128 -0.05876003 -0.06756974   0.03893830
UrbanPop   0.04633575   0.97685748 -0.20054629 -0.05816914
Rape       0.07515550   0.20071807   0.97408059   0.07232502
```

We immediately note that the solution is not the same as when the correlation matrix was analyzed. The resulting scree plot highlights this difference even more. We see below that the first component extracted accounts for much more variance in the unstandardized solution.

```
> scree.cov <- fviz_eig(usa.comp.cov)
> scree.cov
```

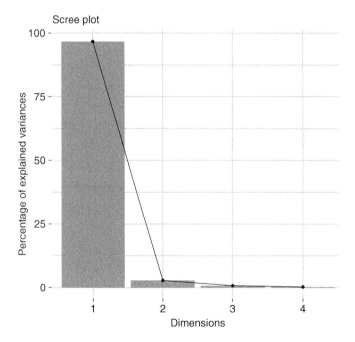

So what's the answer? To standardize or not? Like so many questions in statistics and mathematical modeling in general, there is usually no definitive answer. The safest route is probably to standardize, but as emphasized by Rencher, the covariance matrix is probably more true to the spirit of PCA. If in doubt, there is nothing wrong with running **both solutions**, comparing them, and allowing your readers or audience to see the differences. Above all, the question of whether or not to standardize is pedagogically meaningful, because it highlights, in a strong sense, how such decisions can have huge impacts on results. It serves to emphasize that when it comes to statistics and data analysis, what you get often depends on what you go in with. It's not quite the same as an archeological dig, for example, and should never be treated as such. **The instrument you use to do the digging often influences the nature of what is found. Rarely can a statistical model, by itself, uniquely characterize reality. Statistical analysis does not discover "truth," it merely transforms, translates, and summarizes data subject to the constraints we impose on it**.

It is also one reason why no **single analysis** will provide you with complete definitive insight. The issues of whether to standardize or not, though of course a technical issue, quickly become philosophical as well, in that if the input matters so much to the output or what is "discovered," then it would seem that independent criteria should exist for determining the input. However, across most statistical methods, such independent criteria often do not exist. A similar issue arises, as we will see, in cluster analysis, in which choice of **distance measure** can have rather dramatic effects on the ensuing cluster solution, and often there is no definitive recommendation over which distance measure to use in all situations or all circumstances.

If you do wish to standardize, R offers a convenient way to translate between matrices using the `cov2cor()` function if you wish to specify the correlation rather than covariance matrix directly. As an example, recall the covariance matrix A featured earlier on variables x and y. Below we transform that matrix into a correlation matrix. We name the new matrix R:

```
> A
          x         y
x    6.2662  -3.3811
y   -3.3811   1.9133
```

The corresponding correlation matrix is the following:

```
> R = cov2cor(A)
> R
          x         y
x    1.00000  -0.97648
y   -0.97648   1.00000
```

The **psych** package also has a utility for performing component analysis using the function `principal()`, though not demonstrated here.

Exercises

1 Describe the purpose of principal component analysis, and define what is meant by a principal component. First define it technically, then discuss how it may be useful substantively.

2 Why is it true that PCA will generally generate as many components as there are original variables? Why does this make sense?

3 What does it mean to impose a constraint on components such that their length is equal to 1.0? Why is this constraint relevant and important? What is the consequence if such a constraint is not enforced?

4 Unpack the equation for an eigen analysis. What does the equation for eigenvalues and eigenvectors communicate?

5 Describe how the second component in a PCA is extracted. What are the two conditions it must satisfy?

6 For the component analysis conducted on the USArrests data, attempt to give meaning to the 4 components extracted. Are you able to describe what they might represent?

7 Explain what is being communicated by the following plot (known as a "biplot") generated on the USArrests data of this chapter:

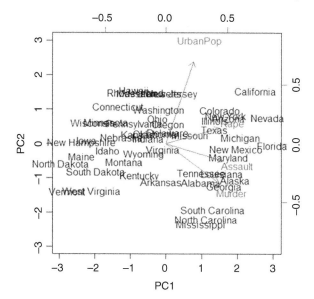

8 Briefly discuss the decision regarding whether to analyze the covariance or correlation matrix in a principal component analysis. When might one solution be more appropriate than the other?

9 Perform a principal component analysis on the iris data's variables `sepal length` through to `petal width` (exclude `Species`), and summarize your findings. How many components were extracted? How many did you decide to keep? Why? Obtain a scree plot of the solution. Does the scree plot help in your decision?

```
> head(iris)
  Sepal.Length  Sepal.Width  Petal.Length  Petal.Width  Species
1          5.1          3.5           1.4          0.2   setosa
2          4.9          3.0           1.4          0.2   setosa
3          4.7          3.2           1.3          0.2   setosa
4          4.6          3.1           1.5          0.2   setosa
5          5.0          3.6           1.4          0.2   setosa
6          5.4          3.9           1.7          0.4   setosa
```

10 Obtain pairwise correlation plots for the variables subjected to the PCA in #9 using the following, and describe both the univariate and bivariate distributions that appear:

```
> pairs()
> library(GGally)
> ggpairs()
```

11

Exploratory Factor Analysis

LEARNING OBJECTIVES

- Understand the nature of exploratory factor analysis (EFA), and how it differs from principal component analysis (PCA).
- Understand the common factor analysis model, and its technical components.
- Understand the major technical and philosophical pitfall of EFA.
- Understand how to address the issue of factor retention.
- Understand the nature of factor rotation, and why rotation is permissible in factor analysis.
- Understand how to plot the results of factor analysis.
- Understand methods of estimation in EFA.
- Perform a complete factor analysis and draw appropriate conclusions.

Exploratory factor analysis (EFA) is a dimension reduction technique similar in some respect to principal component surveyed in the previous chapter, though different enough from PCA that the two should not in any real way be considered equivalent. While both in general have the goal of reducing the dimensionality of the data, the EFA model typically hypothesizes **latent dimensions** or **constructs** that underlie observed variables and their correlations. Much has been made about factor analysis since its inception in the early 1900s, and as we will see, due to some fundamental issues with the procedure, it has historically not always been accepted into mainstream science. Historically, though much criticism has been targeted toward factor analysis, much of that criticism may be better targeted toward the **users** of the procedure rather than the procedure itself. In truth, there is nothing wrong with the actual method. However, there is something wrong with a user not understanding it or using it blindly and potentially drawing inappropriate scientific conclusions from it. We will survey some of these controversies in the current chapter, and offer recommendations regarding how to deal with these issues as they arise.

Univariate, Bivariate, and Multivariate Statistics Using R: Quantitative Tools for Data Analysis and Data Science, First Edition. Daniel J. Denis.
© 2020 John Wiley & Sons, Inc. Published 2020 by John Wiley & Sons, Inc.

11.1 Common Factor Analysis Model

Factor analysis is a huge topic, and many large volumes have been written on the subject (e.g. see Mulaik, 2009). In comparison, principal component analysis (PCA) of the former chapter is a relatively simple technique when compared to factor analysis. What makes factor analysis so complicated? What makes factor analysis so complex is mostly the numerous ways of going about the procedure and the many options available to the researcher, of which often there is no clear rule as to which to prefer. And, since different methods can lead to different results and conclusions, the problem of how to even perform factor analysis is always present. Now this isn't only true of factor analysis, it is true of statistical models in general. However, with EFA, the issue is emphasized. **The output and conclusions you draw from a procedure are only as good as the inputs, model assumptions, and other key decisions you make along the way**. With factor analysis, since there are so many options and complexities, the outcome of the procedure for any analysis may be legitimately criticized for not being remotely "objective." Indeed, factor analysis has often been viewed as half art, half science.

Though there are many models of factor analysis, the most popular historically is what is called the **common factor analysis model**:

$$\mathbf{x} = \boldsymbol{\mu} + \Lambda\mathbf{f} + \boldsymbol{\varepsilon}$$

We unpack the components of the model:

- $\mathbf{x}$ is a vector of observed random variables, that is, $\mathbf{x}$ contains a set of variables on which we have numerical measurements. Examples include a series of achievement tests, psychometric inventories, measures of intellectual performance, etc.
- $\boldsymbol{\mu}$ is a vector of means for the random variables in the vector $\mathbf{x}$.
- Λ is a matrix of factor loadings akin to regression weights in regression, though for this model, they will be applied to $\mathbf{f}$, which is a vector of unobservable common factors, whereas in the multiple regression context, these were actually observed values on variables.
- $\boldsymbol{\varepsilon}$ is a vector of errors, usually called **specific factors** in the factor analysis paradigm, and represents variation left over and unaccounted for by the common factors in the model.

We see then, that the factor analysis model is, in many respects, similar to that of the multiple regression model of previous chapters. However, the key difference is that in the multiple regression model $\mathbf{y} = \mathbf{x}\boldsymbol{\beta} + \boldsymbol{\varepsilon}$, $\mathbf{y}$ was hypothesized to be a function of observed variables in $\mathbf{x}$, weighted by coefficients in $\boldsymbol{\beta}$. In the factor analysis model, instead of $\mathbf{x\beta}$, we have in its place $\mathbf{\Lambda f}$. That is, the loadings in $\mathbf{\Lambda}$ are on **unobservable factors** rather than observable variables. However, both models contain an error term.

In the mathematical development of the factor analysis model, it turns out that the common factor model featured above holds an implication for how one can reconstruct the covariance or correlation matrix among observed variables. Now, what on earth does this mean? To understand this important feature of factor analysis, we need to understand that in all of its complexities and glory, the ultimate goal of factor analysis on a fundamental technical level is to **reproduce the observed correlation or covariance matrix as closely as possible**.

As an example, consider the following correlation matrix (lower triangular shown only):

```
CORR 1.00000
CORR .343 1.00000
CORR .505 .203 1.00000
CORR .308 .400 .398 1.00000
CORR .693 .187 .303 .205 1.00000
CORR .208 .108 .277 .487 .200 1.00000
CORR .400 .386 .286 .385 .311 .432 1.00000
CORR .455 .385 .167 .465 .485 .310 .365 1.00000
```

The goal of factor analysis at a technical level is to reproduce this observed correlation matrix as best as it can. But how does the factor analysis procedure attempt to do this? It does so by choosing weights that as closely as possible regenerate the observed matrix. Though we do not detail the argument here (for details, see Johnson and Wichern, 2007), the factor model above implies the following for the covariance matrix of observed variables:

$$\mathbf{\Sigma} = \mathbf{\Lambda\Lambda'} + \mathbf{\Psi}$$

In this equality, $\mathbf{\Sigma}$ is the covariance matrix of observed variables, $\mathbf{\Lambda}$, as before, are the loadings, and $\mathbf{\Psi}$ is a matrix of specific variances. Pause and recognize the importance and relevance of the above implication. It means that if the factor analysis is maximally successful, then we should, in theory, be able to reproduce the observed covariance or correlation matrix (depending on which matrix we subject to the analysis) based only on the factor loadings in $\mathbf{\Lambda}$ estimated by the factor model, and what is left over will reside in $\mathbf{\Psi}$. Recall in regression analysis the job was to estimate coefficients such that we could minimize the sum of squared errors. Virtually all statistical models try to **maximize**, **minimize**, or **reproduce** something subject to particular constraints. This is a unifying principle to take away from how statistical models work in general.

There are many assumptions and constraints associated with the EFA model. Most are relatively technical and beyond our discussion here. Johnson and Wichern (2007) does a good job of summarizing the assumptions made by the factor model. However, one assumption that is worth emphasizing regarding the common factor analysis model is that it derives factors that are **orthogonal** (Izenman, 2008), which, as was the case for PCA, can be thought of as **uncorrelated** (though at a technical level, orthogonal and uncorrelated do not mean the same thing, it does no harm to think of them as similar in this case). That is, when factors are extracted, each factor is accounting for variance unaccounted for by other factors. If the factor analyst decides to allow factors to correlate, then the solution is called an **oblique** solution.

As was the case for PCA of the previous chapter, there should also be sufficient correlation in the correlation matrix subjected to a factor analysis to make it worthwhile. As was done for PCA then, **Bartlett's test of sphericity** and the **Kaiser–Meyer–Olkin (KMO) measure of sampling adequacy** can be obtained using the **rela** package to evaluate the condition of sufficient correlation and suitability of factor analysis. However, as we warned in our chapter on PCA, these measures should only be employed as crude indicators, and should not in themselves "halt" an EFA from going forth. If there is insufficient correlation in the matrix, then the ensuing factor analysis may simply not be very informative. That is, we certainly do not require the "approval" of Bartlett's before trying the EFA. In the same way, it would be incorrect to stop the factor analysis if KMO was a bit lower than we would have liked (values of 0.7 and higher are generally preferred). Recall that the KMO test attempts to estimate the proportion of variance in the observed variables that may be due to underlying factors; hence, if KMO is quite low, the resulting factor analysis may not reveal a good solution. However, that certainly doesn't mean the EFA cannot be performed. Indeed, after investing in collecting a substantial data set, one should not cease the factor analysis simply because KMO does not meet a given threshold. As was true for Bartlett's, this is not a true test that must be "passed" before moving on to performing the EFA. Use these tests as guidelines and indicators only.

11.2   A Technical and Philosophical Pitfall of EFA

One of the primary criticisms of EFA is that in the aforementioned model, the loadings in Λ are not **uniquely** defined. But what does this mean? First off, "uniquely" in this sense essentially means you will always obtain the same answer regardless of conditions or constraints. However, the values for the loadings in EFA will change as a function of how many factors one chooses to extract from the procedure. That is, if we chose to extract a two-factor solution, the loadings on both factors would be different than if we chose to extract a third factor as well. This is what we mean by saying the loadings are not unique. **They are subject to change on a given factor depending on how many other factors are extracted along with it**. This does not happen in PCA. In PCA, whether the user extracts 2 or 3 (or more) components does not change the loadings or weights associated with those components extracted. Hence, a 2-component solution in PCA will have associated with it the same weights as if we extracted a 3 or 4 component solution.

So why is this a problem for EFA? Why is the non-uniqueness of loadings such a big deal? It's a big deal because when we perform a factor analysis, and find, for instance, that the observed variable quantitative (as one example) loads highly on the IQ factor, it is somewhat disconcerting to us to find had we extracted a different number of factors, the loading for quantitative would be different, and potentially not load as highly on IQ. In some cases, the loadings can change rather drastically, and so EFA carries with it a bit of the impression that what you "find" in the procedure can depend greatly on how many factors you are seeking to find. Hence part of the subjectivity inherent in factor analysis and why some criticize the procedure as revealing what the researcher is wanting to see, rather than anything "objective" in the scientific sense. We will demonstrate using real data the non-uniqueness of loadings issue next. For a deeper discussion and proof of the issue, the interested reader is encouraged to consult Johnson and Wichern (2007) or Rencher (1998) to survey the technical justification behind this argument.

11.3 Factor Analysis Versus Principal Component Analysis on the Same Data

11.3.1 Demonstrating the Non-Uniqueness Issue

To demonstrate the non-uniqueness of loadings issue with factor analysis, we will simultaneously analyze a correlation matrix using PCA of the previous chapter. Since, as we learned, there are typically always as many components generated as there are variables (assuming a well-behaved matrix), this will result in p components. In the EFA, we will extract first a 2-factor solution, then a 3- and 4-factor solution, and compare our results to the component analysis. Recall `cormatrix`:

```
> cormatrix
        c1    c2    c3    c4    c5    c6    c7    c8
[1,]  1.000 0.343 0.505 0.308 0.693 0.208 0.400 0.455
[2,]  0.343 1.000 0.203 0.400 0.187 0.108 0.386 0.385
[3,]  0.505 0.203 1.000 0.398 0.303 0.277 0.286 0.167
[4,]  0.308 0.400 0.398 1.000 0.205 0.487 0.385 0.465
[5,]  0.693 0.187 0.303 0.205 1.000 0.200 0.311 0.485
[6,]  0.208 0.108 0.277 0.487 0.200 1.000 0.432 0.310
[7,]  0.400 0.386 0.286 0.385 0.311 0.432 1.000 0.365
[8,]  0.455 0.385 0.167 0.465 0.485 0.310 0.365 1.000
```

For now, since we are merely wanting to demonstrate the non-uniqueness issue, we present only the loadings of the EFA, and reserve interpretation of other EFA output to later in the chapter when we analyze the Holzinger and Swineford data. We extract a 2-factor solution on `cormatrix`, then a 3-factor, then 4-factor solution:

```
> efa.2 <- factanal(covmat = cormatrix, factors = 2, n.obs = 1000)
> efa.2
> efa.3 <- factanal(covmat = cormatrix, factors = 3, n.obs = 1000)
> efa.3
> efa.4 <- factanal(covmat = cormatrix, factors = 4, n.obs = 1000)
> efa.4
```

```
Loadings:                 Loadings:                          Loadings:
      Factor1 Factor2           Factor1 Factor2 Factor3           Factor1 Factor2 Factor3 Factor4
[1,]  0.978   0.199     [1,]  0.960   0.214   0.165     [1,]  0.661   0.320           0.333
[2,]  0.265   0.420     [2,]  0.235   0.368   0.231     [2,]  0.113   0.699
[3,]  0.447   0.340     [3,]  0.441   0.448             [3,]  0.196   0.145   0.144   0.956
[4,]  0.153   0.795     [4,]  0.110   0.780   0.211     [4,]          0.542   0.389   0.257
[5,]  0.673   0.178     [5,]  0.643   0.107   0.320     [5,]  0.939           0.108
[6,]          0.591     [6,]          0.582   0.122     [6,]          0.131   0.979   0.104
[7,]  0.307   0.502     [7,]  0.284   0.474   0.161     [7,]  0.237   0.441   0.346   0.132
[8,]  0.365   0.500     [8,]  0.247   0.313   0.914     [8,]  0.436   0.523   0.208
```

We remark on differences between the three solutions:

- We see that the loadings on the 2-factor solution do not carry over to the 3-factor and 4-factor solutions. For instance, for T1 in the 2-factor solution, the loading for factor 1 is equal to 0.978, while for T1 in the 3-factor solution the loading for factor 1 is equal to 0.960. Though in this case the change is small, it is still apparent that the loading did change simply as a result of extracting an additional factor. Likewise, the loading on factor 2 changed from 0.199 to 0.214.
- The change in loadings is even more dramatic when we consider the 4-factor solution. For T1, factor 1, notice the loading dropped from 0.978 in the 2-factor solution to 0.661 in the 4-factor solution.

- Further evidence that loadings can change rather substantially depending on which solution is extracted is evident by comparing the 2-factor solution to the 3- and 4-factor solutions. Especially remarkable, for example, is the loading for T5 on factor 1 in the two-factor solution, equal to 0.673, then increasing to 0.939 in the 4-factor solution. Changes such as these are quite substantial changes and non-negligible, and is a clear demonstration of what we mean when we say loadings are not unique across factor solutions. That is, **how well a given variable loads onto a given factor often depends on how many factors were extracted in the factor analysis**.

Now, suppose instead of EFA, we conducted a PCA on the same data:

```
> pca <- princomp(covmat = cormatrix)
> loadings(pca)
```

```
Loadings:
    Comp.1 Comp.2 Comp.3 Comp.4 Comp.5 Comp.6 Comp.7 Comp.8
c1   0.413  0.458                        0.135         0.756
c2   0.303 -0.114 -0.637  0.471          0.470 -0.132 -0.144
c3   0.318         0.546  0.581 -0.122 -0.192 -0.368 -0.265
c4   0.373 -0.430                0.111 -0.471 -0.176  0.639
c5   0.357  0.543               -0.314         0.219  0.311 -0.574
c6   0.301 -0.493  0.381 -0.401                0.541 -0.245
c7   0.366 -0.216                0.775 -0.444  0.105
c8   0.381        -0.364 -0.396 -0.382 -0.395 -0.512
```

In this component analysis of the data, succinctly put, **the component loadings are what the component loadings are, period. That is, unlike factor analysis, they are not amenable to change depending on how many components are "kept" in the procedure**. In component analysis, there is no analogue to requesting a 2- or 3- or more-component solution. Yes, we are free to **keep** only those components that we deem meaningful, but the point is that PCA typically (for well-behaved matrices) generates as many components as there are variables, and whether we decide to retain (keep) the first component only or more, the loadings generated by the procedure will not change.

Hence, reading off the above output for PCA, we can make conclusions of the sort: "T1 loads on component 1 to a degree of 0.413," without mentioning how many components we chose to extract in the procedure. Again, that is because the weights (loadings) associated with the components are stable. In EFA, however, a statement such as that we just made is uninformative unless we include mention of the context in which the loading occurred. That is, in making the analogous interpretation in EFA, we would have to add the qualifier, " ... given an n-factor solution was extracted." Again, this is because the weights in factor analysis are subject to change depending on how many factors we extract. For this reason, it is probably better to reserve the term "extract" when referring to specifying the number in a factor analysis, whereas in PCA, it is better to refer to this as the number of components "kept."

When conducting an EFA, the loading sizes (and even directions in some cases) will likely change depending on how many factors are specified for extraction. This is part of the bigger problem regarding the non-uniqueness of loadings issue in exploratory factor analysis. In principal component analysis, loadings are stable, in that regardless of how many components are kept, the loadings do not change. For this reason, it is probably better to refer to the number of factors requested in factor analysis as the number "extracted," whereas in PCA, it is better to refer to the number of components you are retaining as the number "kept."

11.4   The Issue of Factor Retention

With today's software capabilities, virtually anyone, after reading an introductory chapter on the topic, or even just digging up some software code, can perform a factor analysis in a matter of milliseconds. Both the blessing and the curse of modern software is that most statistical analyses can be performed at lightning speed. However, that in no way necessarily correlates to how interpretable the solution will be or whether the factor analysis should have even been performed in the first place. This is true of all statistical methods, but perhaps no more true than in factor analysis due to its already inherent subjective nature.

The decisions facing the factor analyst, aside from the myriad of technological input decisions such as type of factor analysis, type of rotation, etc. include the following:

- How many factors will I ask R to extract for a given factor analysis run? Given that the loadings are not unique across factor solutions, the question is how many factors should be specified for the program to "pull out" of the data?
- Given that the number of desired factors are extracted, how many of those factors should I retain? For instance, if I decide to extract a 3-factor solution, how many of these factors can I actually make sense out of and thus retain in my final solution?

- If deciding on retaining 2 factors out of the 3 extracted, should I revert back to asking R to conduct a 2-factor solution to begin with, and hope the loadings stay more or less stable and consistent?

None of these questions are easy to answer, largely because they have no definitive answers! There is no clear road map or decision tree to conduct exploratory factor analyses, but at minimum, you should be ethical and report the **process you followed** so that your reader or audience can then interpret your steps accordingly. Factor analysis is as much about **process** as it is about **product**. Often in statistical methods, one cannot properly evaluate output without an understanding of what went through the mind of the investigator in generating the output. This is especially true with regard to factor analysis.

11.5 Initial Eigenvalues in Factor Analysis

In beginning a factor analysis, it is usually convention to start by getting an idea of the magnitude of the eigenvalues based on a principal component solution. As mentioned, since the loadings in PCA are stable and do not depend on the number of components specified in a solution, they give a nice precursor inspection to doing the factor analysis. However, as always, it should be kept in mind that the PCA is not a factor analysis, and hence the magnitudes of the eigenvalues should be used as a preliminary guide only to working toward deciding on the ultimate number of factors to retain in a solution.

In R, we obtain the eigenvalues using `eigen()` on `cormatrix`. Since we are only interested in the eigenvalues, we do not include output about eigenvectors (R will automatically produce the 8 eigenvectors as well, each corresponding to each of the eigenvalues):

```
> eigen(cormatrix)

eigen() decomposition
$`values`
[1] 3.4470520 1.1572358 0.9436513 0.8189869 0.6580753
0.3898612 0.3360577
[8] 0.2490798
```

We then obtain a plot of the eigenvalues as follows:

```
> pca <- princomp(covmat = cormatrix)
> plot(pca, type = "lines")
```

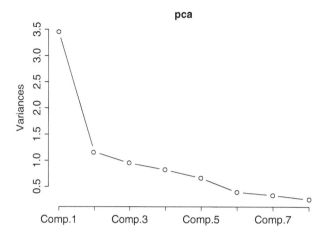

The plot generated is the well-known "scree plot" featured in Chapter 10, and reveals that the first component dominates with a relatively large eigenvalue, while the other components trail off in terms of their magnitudes. Hence, at first glance, maybe a one- or two-factor solution (maybe 3) would make the most sense for these data. Again, however, it is impossible to conclude for sure because we have not actually performed the factor analysis. And what is more, as we have discussed, it is unwise to make a decision on the number of factors to extract until we inspect the derived loadings, possibly subject to rotation.

11.6 Rotation in Exploratory Factor Analysis

As discussed in this chapter already, the primary defining criteria as to whether or not a factor analysis has been "successful" or not is usually by whether or not substantive sense can be made from the factor solution. If the researcher can look at the factors and "see" a sensible factor solution, then the EFA is usually deemed a success, however philosophically absurd that may sound. On the other hand, if the solution makes little sense on a substantive or scientific level, then the factor analysis is usually considered to have failed.

Often, in trying to make sense of a factor solution, researchers may wish to perform a **factor rotation** of the solution. A rotation sometimes can help make better sense of the factor structure and lead to more definition in the obtained factor solution. If a researcher rotates the factors such that they remain uncorrelated, then it is referred to as an **orthogonal rotation**. If the researcher rotates the factors such that they are allowed to correlate, then it is called an **oblique rotation**. Whether one performs an orthogonal or oblique rotation often depends again on which solution is more interpretable and whether having uncorrelated or correlated factors makes more substantive sense to the researcher and scientific community.

The most common method of orthogonal rotation is, by far, **varimax rotation**. In a varimax rotation, the goal is to maximize within-factor variance of the loadings of those factors extracted. But what does this mean, exactly? To understand this, consider once more the two-factor solution obtained earlier:

```
Loadings:
       Factor1  Factor2
[1,]   0.978    0.199
[2,]   0.265    0.420
[3,]   0.447    0.340
[4,]   0.153    0.795
[5,]   0.673    0.178
[6,]            0.591
[7,]   0.307    0.502
[8,]   0.365    0.500
```

Looking at the loadings for factor 1, we see a distribution of loadings ranging from very small (so small that the loading for T6 was omitted by R), to quite large (0.978). We could, if we wanted to, compute the variance of the loadings. This is the amount of spread or variability we have in the column of loadings for factor 1. What a varimax rotation seeks to accomplish is to **maximize this variance**. But how can this variance be maximized? It can be maximized through **driving larger loadings to be even larger, and smaller loadings to be even smaller**. That is, varimax rotation will try to more or less draw a line in the sand about which variables load high versus which are not worth looking at very much. Varimax rotation is so common in factor analysis, that the factanal() function we just ran in R for these data included it automatically. Hence, the loadings we obtained, by default, are varimax loadings. The following generates the same output that we obtained for the 2-factor solution, where this time we explicitly designate the varimax rotation:

```
> efa.2 <- factanal(covmat = cormatrix, factors = 2, n.obs
= 1000, rotation = "varimax")
```

We had mentioned that sometimes the researcher wishes to allow factors to correlate in the rotation. Remember that by definition, orthogonal factor analysis generates factors that are uncorrelated, and the varimax rotation keeps factors uncorrelated, or orthogonal at a 90-degree angle. For details on oblique rotational methods in R, refer to the **psych** package.

The type of factor analysis one conducts, including which rotation is performed, can have small to dramatic effects on the final factor analysis solution in terms of the size and distribution of the estimated loadings on each factor. Hence, not only will the number of factors the user chooses to extract help determine the make-up of the factors, but the rotational method (e.g. varimax, promax) chosen can likewise generate drastic differences. All good factor analysts and interpreters of factor analysis results need to keep these things in mind.

11.7 Estimation in Factor Analysis

When we conduct a factor analysis, as is true of virtually all statistical methods, parameters are being estimated. That is, we require statistical estimation to estimate such things as the loadings and the communalities. Estimation methods in factor analysis essentially define the procedure. For instance, factanal() uses **maximum likelihood** as its **estimator**. So if a colleague asked you what type of factor analysis factanal() conducts, your answer would be that it conducts **maximum likelihood factor analysis**. Maximum likelihood EFA is a very common version of factor analysis, and in most cases, is the preferred choice. However, other methods exist, including **principal axis factoring**. In principal axis factoring, or **paf** for short, the algorithm provides initial communality estimates (crudely, an estimate of how much variance the given variable has in common with other variables in the set) in the form of the squared multiple R from regressing the given variable in the factor analysis on the set of remaining variables. For example, for T1, we could obtain an initial communality estimate by regressing T1 onto the remaining variables (T2 through to T8), then use this estimate as an "initial communality" to essentially give the factor analysis algorithm something to start in terms of how much communality the given variable shares with remaining variables. Then, once the iteration on the solution begins, the algorithm will "polish up" these initial estimates and generate final communalities. Principal axis factoring can be accomplished using the **psych** package in R using the factor.pa() function.

11.8 Example of Factor Analysis on the Holzinger and Swineford Data

We demonstrate an EFA using the 9 tests from the Holzinger and Swineford data. These data consist of mental ability tests on groups of children. Though the original Holzinger and Swineford data consists of more than 9 tests, our example features only tests of visual perception (`vis_perc`) through to speeded discrimination straight and curved capitals (`s_c_caps`). First, we obtain the correlation matrix of the 9 tests:

```
> library(psych)
> Holzinger.9
            vis_perc    cubes lozenges par_comp sen_comp wordmean addition
vis_perc    1.00000 0.325800 0.448640  0.34163  0.30910  0.31713 0.104190
cubes       0.32580 1.000000 0.417010  0.22800  0.15948  0.19465 0.066362
lozenges    0.44864 0.417010 1.000000  0.32795  0.28685  0.34727 0.074638
par_comp    0.34163 0.228000 0.327950  1.00000  0.71861  0.71447 0.208850
sen_comp    0.30910 0.159480 0.286850  0.71861  1.00000  0.68528 0.253860
wordmean    0.31713 0.194650 0.347270  0.71447  0.68528  1.00000 0.178660
addition    0.10419 0.066362 0.074638  0.20885  0.25386  0.17866 1.000000
count_dot   0.30760 0.167960 0.238570  0.10381  0.19784  0.12114 0.587060
s_c_caps    0.48683 0.247860 0.372580  0.31444  0.35560  0.27177 0.418310
            count_dot s_c_caps
vis_perc      0.30760  0.48683
cubes         0.16796  0.24786
lozenges      0.23857  0.37258
par_comp      0.10381  0.31444
sen_comp      0.19784  0.35560
wordmean      0.12114  0.27177
addition      0.58706  0.41831
count_dot     1.00000  0.52835
s_c_caps      0.52835  1.00000
```

To familiarize ourselves with the matrix, we interpret a few of the correlations:

- We can see that values of 1.00000 lie along the main diagonal of the matrix. This is how it should be, since it is a correlation matrix, and a variable correlated with itself is equal to 1. For instance, the first variable, visual perception (`vis_perc`) has a correlation with itself equal to 1.
- The correlation between visual perception (`vis_perc`) and cubes is equal to 0.325800.
- The correlation between lozenges and cubes is equal to 0.417010.
- The correlation between counting dots (`count_dot`) and addition is relatively high compared to the rest of the correlations at a value of 0.587060.

We can also visualize the magnitudes of correlation using cor.plot() by the degree to which cells are shaded. The first plot below does not contain the numerical correlations within each cell. The second plot does, by the addition of numbers=TRUE:

```
> library(psych)
> cor.plot(Holzinger.9,numbers=TRUE)
```

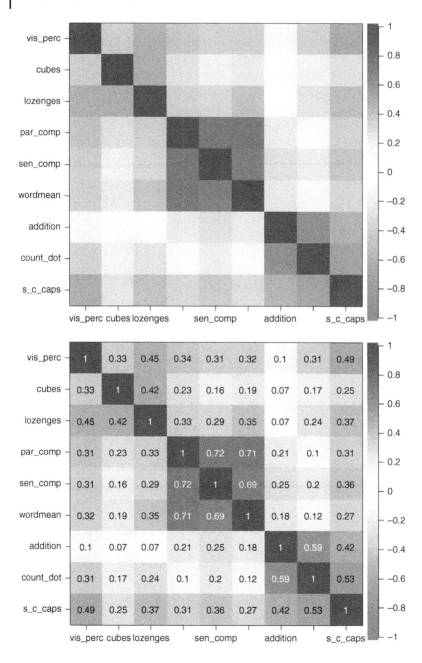

We now proceed with requesting a factor analysis in R:

```
> library(psych)
> fa <- factanal(covmat = Holzinger.9, factors = 2, n.obs = 145,
rotation = "varimax")
> fa
```

We are naming the object `fa`, and using `factanal` in the **psych** package. This will generate for us a factor analysis based on **maximum likelihood**. To reiterate how the function works, we specify the matrix by `covmat = Holzinger.9` and the number of factors we wish to extract by `factors = 2`. The number of observations on which the matrix is based is indicated by `n.obs = 145`, and we are requesting a varimax rotation (`rotation = "varimax"`).

```
Call:
factanal(factors = 2, covmat = Holzinger.9, n.obs = 145,
rotation = "varimax")

Uniquenesses:
vis_perc   cubes lozenges par_comp sen_comp wordmean addition
   0.733   0.899    0.781    0.237    0.327    0.323    0.595
count_dot   s_c_caps
    0.253      0.514

Loadings:
          Factor1 Factor2
vis_perc   0.354   0.376
cubes      0.232   0.219
lozenges   0.364   0.293
par_comp   0.866   0.112
sen_comp   0.794   0.205
wordmean   0.815   0.114
addition   0.126   0.624
count_dot          0.864
s_c_caps   0.288   0.635

               Factor1 Factor2
SS loadings      2.455   1.882
Proportion Var   0.273   0.209
Cumulative Var   0.273   0.482

Test of the hypothesis that 2 factors are sufficient.
The chi square statistic is 61.7 on 19 degrees of freedom.
The p-value is 2.08e-06
```

We interpret the output:

- The first piece of output is simply restating the input to R we requested:

```
Call: factanal(factors = 2, covmat = Holzinger.9, n.obs =
145, rotation = "varimax")
```

- Next are the Uniquenesses: These correspond to the specific variances for each variable, that is, that portion of variance unexplained by the factor solution. For example, the first number, that of 0.733, was obtained by inspecting the loadings for vis_perc, squaring and summing them, which gives us $0.354 \times 0.354 + 0.376 \times 0.376 = 0.125316 + 0.141376 = 0.266692$. The number 0.266692 is called the **communality** for visual perception. The number 0.733 is what is left over, so that the explained variance and the unexplained variance sum to 1.0. That is, $0.266692 + 0.733 = 1.0$ (within rounding error). A high uniqueness value suggests the factor solution isn't doing a great job at explaining that particular variable.

- The loadings (Loadings) are given for each factor: factor 1 and factor 2. These reveal the "structure" of each factor. For instance, par_comp loads quite high on factor 1 (0.866) and count_dot loads quite high on factor 2 (0.864).

- The loading for count_dot on factor 1 is not shown because it is quite small, and R automatically suppresses it. We can uncover that loading by requesting a much smaller cut-off in R (try > print(fa, digits = 2, cutoff = 0.001, sort = FALSE, where we are setting the cutoff at 0.001, which is quite small. You will see that the loading of 0.01 for count_dot is now revealed).

- SS loadings corresponds to the sum of squared loadings for each factor. For instance, for factor 2, we can demonstrate the computation:

```
> factor.2 = 0.376*0.376 + 0.219*0.219 + 0.293*0.293 +
0.112*0.112 + 0.205*0.205 + 0.114*0.114 + 0.624*0.624 +
0.864*0.864 + 0.635*0.635
> factor.2
[1] 1.881848
```

- Proportion Var is the proportion of variance accounted for by each factor. This is computed as SS loadings for the given factor divided by the number of observed variables subjected to the factor analysis, which in this case, is 9. Since the SS loadings for each factor are 2.455 and 1.882, respectively, the proportion of variance accounted for by factor 1 is $2.455/9 = 0.2728$, while the proportion of variance accounted for by factor 2 is $1.882/9 = 0.209$. These numbers of 2.455 and 1.882 are the "new eigenvalues" based on the factor solution (as opposed to the "old" ones first generated by the principal component solution). It's a bit "risky" to call these by the name of eigenvalues in this sense (as on a technical level, it confuses with PCA, and they are not actually eigenvalues). However, so long as one understands which values correspond to which analysis, it does no great harm to think of them this way for applied purposes. They provide us with the proportion of variance explained by the factors, which is similar in concept to the variance of the principal components (which **are**, properly speaking, the **original eigenvalues** of the matrix, that is, of the eigen decomposition for component analysis).

- `Cumulative Var` is simply the sum of `Proportion Var` explained by each factor. That is, for our data, it is equal to $0.2728 + 0.209 = 0.482$. Hence, for our data, these two factors account for approximately 48% of the variance.
- Inferential tests for factor analysis are often not performed or given in software, but `factanal` does provide us with such a test based on a chi-squared distribution. The null hypothesis is that 2 factors are sufficient to account for these data. Since the obtained p-value is extremely small ($p = 2.08e-06$), we reject the null and can conclude that a two-factor solution is not sufficient. However, inferential tests of this manner are not the be-all or end-all of factor analysis, and these tests should be interpreted with caution. Much more important to the factor analyst will be whether the obtained factor structure makes **substantive sense**, rather than the results of an inferential test. Hence, though we report this test, we do not base our final factor analysis conclusions on it alone.

11.8.1 Obtaining Initial Eigenvalues

In the above output, we had said the values of 2.455 and 1.882 for `SS loadings` were obtained based on the factor analysis, with the caveat that they were not actual eigenvalues as is the case of principal components based on an original eigen decomposition. So why mention eigenvalues at all here then? As previously discussed, the reason is because prior to conducting a factor analysis, researchers are sometimes interested in the **original eigenvalues** based on the data before it was subjected to a factor analysis. In software such as SPSS, the PCA is generated automatically. These "initial" eigenvalues then, are based on a PCA. We can obtain these eigenvalues in R by the following:

```
> ev <- eigen(Holzinger.9)
> ev

eigen() decomposition
$values
[1] 3.6146529 1.5622792 1.2505174 0.7067834 0.5327174
0.4213225 0.3580553 0.2979833 0.2556886
```

The above then is the original eigen decomposition performed on our data. That is, it is the result of a PCA. These "initial eigenvalues" can be used as a general guideline to help determine the number of factors, so long as it is understood that only the final factor solution will yield the true variances explained by the factors. **The PCA alone is not a verdict on the final factor solution**. In SPSS software, for instance, the output encourages the user to see the SS loadings as "updated" from the original PCA eigenvalues. The "update" of course, is based on conducting the factor extraction instead of merely the PCA.

11.8.2 Making Sense of the Factor Solution

The most important question for the researcher will be whether the obtained factor solution makes any sense. We reproduce the loadings below:

```
Loadings:
          Factor1 Factor2
vis_perc  0.354   0.376
cubes     0.232   0.219
lozenges  0.364   0.293
par_comp  0.866   0.112
sen_comp  0.794   0.205
wordmean  0.815   0.114
addition  0.126   0.624
count_dot         0.864
s_c_caps  0.288   0.635
```

Inspecting the loadings, we notice that paragraph comprehension (par_comp), sentence comprehension (sen_comp), and word meaning (wordmean) appear to load quite heavily on factor 1, while addition (addition), counting dots (count_dot), and sc caps (s_c_caps) load fairly high on factor 2. Hence, we might tentatively conclude that the first factor corresponds to a **verbal ability** (or some such construct), while the second a **quantitative ability**. Again, however, the factor analysis by itself cannot confirm the naming of the factors. As latent variables, we are free to name them whatever makes the most sense to us and to the research community as a whole. **As with any variable, it is only as real as we, as a scientific community, agree it is**. We can also obtain the loadings for each factor via fa$loadings[,1:2].

We generate the corresponding plot in R using cluster.plot() to visualize the loadings in bivariate space across both factors:

```
> library(psych)
> cluster.plot(fa)
```

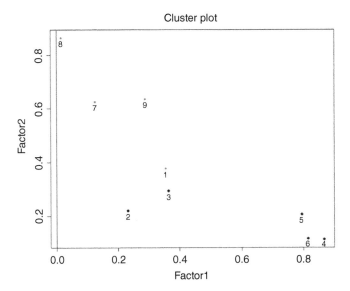

Cluster plot

The numbers in the plot correspond to the given variable in our data. For instance, variables 4 and 6 (par_comp and wordmean) load highly on factor 1 (bottom right corner of the plot) and low on factor 2. Variable 8 (count_dot) loads very low on factor 1, but very high on factor 2 (top left corner of the plot).

Exercises

1 From both technical and substantive points of view, discuss the goals of exploratory factor analysis (EFA).

2 How is the EFA model different from that of the regression model? What is the key difference?

3 What is the primary technical and philosophical pitfall of EFA? How is this a significant issue when it comes to the extraction and interpretation of factors?

4 Your colleague conducts a factor analysis, and concludes to you, "I just found two factors in my analysis." List a few questions you would like to ask her as follow-ups to her statement, as to learn more about the strength of her claim.

5 Why is the idea of "keeping" two components in PCA different from that of keeping two factors in EFA? Discuss.

6   Your colleague says to you, "Rotating factor solutions is akin to cheating and 'massaging' the solution." Do you agree? Why or why not?

7   Consider once more the Holzinger.9 data analyzed in this chapter, but this time, drop the variable vis_perc from the solution and conduct the EFA anew requesting once more a 2-factor solution. Summarize your findings and comment on any differences you observe between this solution and that obtained in the chapter. Why are these general observations that you're making important in the bigger context of what factor analysis accomplishes?

8   Using the factanal() function, report the *p*-value obtained in the analysis conducted in #7, and comment on the degree to which it is different from the *p*-value obtained on the original analysis in the chapter. Why would it be different?

9   Plot the solution obtained in #7 using cluster.plot(), and briefly describe what the plot represents.

10   After reading on cluster analysis in the following chapter in this book, briefly discuss how factor analysis and cluster analysis are similar. How are they different?

12

Cluster Analysis

LEARNING OBJECTIVES

- Understand the nature of how cluster analysis works, and appreciate why it is sometimes called an "unsupervised learning" technique in the machine and statistical learning literature.
- Through the simplest of examples, obtain a strong understanding of what cluster analysis actually accomplishes (and perhaps more importantly, what it does not accomplish).
- Understand the distinction between the algorithm determining the existence of clusters and the researcher confirming them.
- Understand that cluster analysis is determined by measures of distance, the type of which can be chosen at the liberty of the researcher.
- Distinguish between k-means clustering and hierarchical clustering.
- How to interpret a dendrogram in hierarchical clustering.
- Why clustering is inherently subjective and why validating a solution can prove challenging.

In this chapter, we survey the statistical method of **cluster analysis**, and provide demonstrations of how to perform the procedure in R. Recall that in factor analysis, we were primarily interested in learning whether variables might "go together" in some sense to uncover latent dimensions (or "factors"). In cluster analysis, we are primarily interested in learning whether cases (people or other units of observation) can be lumped together in a sense such that they generate more or less "natural" groupings or classifications.

As an example of a situation that may call for a cluster analysis, suppose we give a psychological inventory to a sample of 10,000 individuals, consisting of a battery of psychological tests. Suppose the inventory measures such things as depression

Univariate, Bivariate, and Multivariate Statistics Using R: Quantitative Tools for Data Analysis and Data Science, First Edition. Daniel J. Denis.
© 2020 John Wiley & Sons, Inc. Published 2020 by John Wiley & Sons, Inc.

symptomology, anxiety, psychosis, and so forth. After collecting data on the 10,000 units, we ask the following question:

Are there any naturally occurring groups that could be extracted from these data?

By "naturally occurring," it is meant to imply groups that in some sense, "exist" in the data, and that we would like to uncover and potentially learn more about. That is, imagine that after studying these data, we come to the conclusion that there are two primary groups:

1) those suffering from psychological abnormality or distress
2) those not suffering

Imagine now that we had a statistical algorithm capable of uncovering these groups and allowing their natural structure to surface, such that after applying the algorithm, we could draw the conclusion that the grouping structure does indeed exist. The technique of cluster analysis is a statistical method designed to search and potentially find such clusters or groups in the data, that, not unlike factor analysis, may or may not make substantive sense to the researcher conducting the analysis.

However, it must be emphasized from the outset that cluster analysis will not necessarily "discover" on its own, **natural** groupings or clusters. Rather, as we will see, it will simply apply an algorithm to data that facilitates separation between potentially substantively meaningful clusters. And how those clusters are defined will depend often on what metric we use to assess distance between cases. Hence, whether or not cluster solutions are successful, even at the mathematical level, will often depend on what metric we use to define similarity (or dissimilarity) among cases.

In the **machine learning** and **statistical learning** literature, cluster analysis is usually designated as an **unsupervised learning** technique. Calling it "unsupervised" is just fancy jargon for saying that when we apply the technique, we beforehand do not have any already **predefined structure** to the data. That is, we are allowing the algorithm to do the search and separation for us with the hope that a grouping structure will reveal itself that will make some substantive sense to the researcher. Again, however, even if group separation is found, as will usually be the case with any cluster analysis you do, it does not necessarily mean the grouping structure will make any immediate (or even long-term) substantive sense. It is entirely possible that the grouping structure found is, and forever will be, **meaningless**.

The issue is somewhat philosophical, but is important, because it begs the deeper question that should cluster analysis find groupings, yet we cannot immediately identify what those groupings are, does it mean the cluster analysis was all for naught, or does it suggest our science simply isn't matured or developed enough for us to know what these groupings actually represent substantively aside from their mathematical determination? Of course, that issue is far (far) beyond the scope of this book, but is a substantive question you may have to confront if you perform cluster analysis and present it to a critical audience. For further details on these philosophical issues, see Hennig (2015). Realize, however, that upon consulting papers such as this, you will be initiated not to further statistical methods, but rather to the **philosophy of statistics** and science, which in itself is a marvelous field of inquiry. Indeed, digging into some (or all) of such issues is necessary if you are to intelligently situate where statistical methods in general belong in the general quest for knowledge, scientific or otherwise.

12.1 A Simple Example of Cluster Analysis

As we have repeatedly seen throughout this book, understanding any statistical technique is greatly facilitated by considering exceptionally simple examples on very simple data. Then, when you encounter more complex situations, the increase in complexity isn't perceived as something categorically "different" than the simple example, but rather is simply an extension of the simplicity! Remember, to so-called "geniuses," things are (or eventually become) **simple**. It is only to us non-geniuses that things are complex. Always seek to reduce **perceived complexity** down to its most basic, elementary principles, and such will instill new perception into what was previously interpreted as complex.

To appreciate the nature of cluster analysis, consider the data in Table 12.1 on the heights and weights of four individuals: Mary, Bob, Julie, and Mark.

We summarize the information in the table:

- Mary has a height of 4 ft and 2 in., and a weight of 120 lb.
- Bob has a height of 5 ft and 2 in., and a weight of 190 lb.
- Julie has a height of 5 ft and 8 in., and a weight of 180 lb.
- Mark has a height of 4 ft and 3 in., and a weight of 130 lb.

The question cluster analysis essentially asks of this data is the following:

Given data on height and weight of these four individuals, are there any natural groupings that might be uncovered?

Before considering the mathematics of cluster analysis, let's brainstorm a bit. Look at the data. At first glance, let's entertain the hypothesis that we might be able to separate individuals into whether they are male versus female. That is, could it be that a natural grouping variable to the data is that of gender? Probably not, since we note Mary's height and weight is very similar to that of Mark's. Likewise, Bob and Julie's height and weight are quite similar. Hence,

Table 12.1 Fictional data for simple cluster analysis.

Person	Height	Weight (lb)
Mary	4 ft 2 in.	120
Bob	5 ft 2 in.	190
Julie	5 ft 8 in.	180
Mark	4 ft 3 in.	130

trying to distinguish these cases based on a variable such as gender seems, even on a quick visual level, to be a hopeless venture. Surely, whatever algorithm we subject to these data should not suggest that Mary and Julie belong in one "cluster" while Bob and Mark in another. At least that's our first impression.

What's interesting about cluster analysis is that usually, the clustering variable is not already included in our data set. That is, it's usually a variable **associated** with the cases in our data, but it's not something we can explicitly **see** in our data. To appreciate this point, let's keep looking at the data. What do we notice? We notice that Mary and Mark are very **similar** in measurements on height and weight. That is, both Mary and Mark have heights and weights that are very **close in proximity**. After all, Mary's height of 4 ft, 2 in. is very close to Mark's height of 4 ft, 3 in. Likewise, Mary's weight of 120 lb is very close in proximity to Mark's weight of 130 lb. Their **distances** on height and weight are minimal.

Looking further at the data, we also see that Bob's height and weight are very close in proximity to that of Julie's. We may be onto something. What we are doing by this little exercise is an informal cluster analysis, where we are seeking similarity between cases on the measured variables. This similarity may in the end suggest a grouping structure to the data.

As always, plotting and visualizing the data will likely help immensely. Let's plot the data on weight and height (Figure 12.1).

What do we notice from the plot? We note that points in the lower left of the plot, that is, those persons having low height and low weight, seem to "group" together. Likewise, points in the upper right-hand of the plot also seem to group together. The two circled groups in part (b) of the plot have observations such that within each group, are very similar to one another. That is, the observations share a degree of "within-group" proximity. What we have just identified are two potential "clusters" of observations. Now, why were these clusters formed? Mathematically, simply because observations within each cluster are **similar** on the measurements. However, as a scientist, you wish to conclude more, and will take the results from the cluster analysis and start the speculation as to a **substantive** reason why

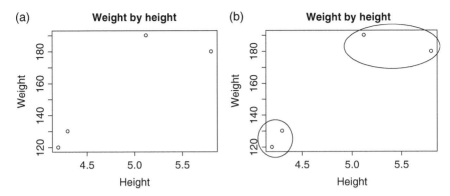

Figure 12.1 (a) Plot of height and weight. (b) Identifying similarity.

the algorithm discovered such separation. For what substantive reason could these two clusters have been formed? You look at your data file, and realize that not only were Mark and Mary of relatively low weight and low height, but that they were also both **children**, while Julie and Bob were both **adults**. That is, the mysterious clustering variable, you conclude, is that of **age classification**, on which for these data there are two categories, child versus adult. Hence, you tentatively conclude that what separates these data points is the age variable.

Though there is of course much more to cluster analysis that the above example would suggest, the "gist" of cluster analysis is really no more complex than that. It's called an **unsupervised learning** technique in the statistical and machine learning literature because presumably, we didn't yet know the **latent structure** of the classes. Instead, we subjected a clustering algorithm for the purpose of, in some sense, separating cases into classes. Then, we inspected the formed groups more closely, and speculated as to why they were different. One possible reason why they were classified as such is their age, but of course, it may not be the only classifier. After all, we could also probably conclude that Mary and Mark both enjoy candy and playing video games more than Bob and Julie, and the cluster analysis would certainly not object to us naming these cluster-analysis-generated categories in this fashion. The key point to remember concerning cluster analysis is the following:

Though a cluster algorithm will usually generate clusters mathematically, the substantive reasons for why those clusters were generated are often many, and no clustering algorithm will reveal, by itself, the key discriminating variable responsible for the clustering. All the cluster algorithm can do is mathematically separate cases based on some distance or proximity metric. The rest is up to the scientist to discover what features are most "responsible" or "relevant" to the cluster separation.

12.2 The Concepts of Proximity and Distance in Cluster Analysis

The key things to understand then in a cluster analysis are the concepts of **proximity** and **distance**. By far the most common way of measuring distance between vectors is the **Euclidean distance**. As an easy example of the Euclidean distance between two vectors, consider the following vectors **x** and **y**:

$$\mathbf{x} = [2, 4]$$
$$\mathbf{y} = [4, 7]$$

To compute the Euclidean distance between these two vectors, we compute $(2-4)^2$ and $(4-7)^2$ and sum these, which gives $4 + 9 = 13$. Then, we take the square root, which yields 3.61. That is, the Euclidean distance between these two vectors is 3.61. The Euclidean distance was obtained by subtracting then squaring the elements in each vector pairwise. That is, the first element in **y** was subtracted from the first element in **x**, while the second element in **y** was subtracted from the second element in **x**, and the square root of the sum of those squared differences was taken.

As a generalization and extension of the above computation, cluster analysis generally computes **distance matrices** containing many cases. What can make cluster analysis relatively complex and confusing is that Euclidean distance is not the only way to define a measure of distance. That may at first sound surprising, but there are other ways of defining distance between two objects. Other possibilities include the **Minkowski metric** and the **city-block** (also known as "Manhattan") measure. We do not detail these distances here, though the interested reader is encouraged to consult Everitt et al. (2001) for more details. It can be difficult to decide among distance measures when conducting a cluster analysis, and so researchers are generally encouraged to try a few different distance measures when performing the procedure, and then observe whether these different ways of defining distance more or less converge on a similar cluster-analytic solution. If they do not, then a deeper inspection of the data and why differences between solutions might be occurring would be in order.

12.3 *k*-Means Cluster Analysis

In **k-means clustering**, the number of presumed clusters must be stated in advance by the researcher. *k*-Means will require that each observation in the data be classified into one of several clusters of which the exact number is determined by the user. To understand *k*-means, we are best left being instructed by one of its innovators, James MacQueen (1967), who described the technique beautifully in an early paper featuring the technique:

> The main purpose of this paper is to describe a process for partitioning an N-dimensional population into k sets on the basis of a sample. The process, which is called "k-means", appears to give partitions which are reasonably efficient in the sense of within-class variance ... State [sic] informally, the k-means procedure consists of simply starting with k groups each of which consists of a single random point, and thereafter adding each new point to the group whose mean the new point is nearest. After a point is added to a group, the mean of that group is adjusted in order to take account of the new point. Thus at each stage the k-means are, in fact, the means of the groups they represent (hence the term k-means). (pp. 281, 283)

A key feature of k-means is that clusters are not allowed to overlap, which means that if a case is in a given cluster, it cannot be in another. That is, cluster membership is mutually exclusive. The procedure begins with k equal to the number of cases in the sample. As noted by MacQueen, after a new case is added to a given cluster, the mean of that cluster is recalculated, thereby incorporating the new data point. Iteratively then, cluster means are updated based on new members to the respective different clusters. These updating cluster means are usually referred to as **cluster centroids**. To initiate the clustering algorithm, initial seeds are randomly chosen for each cluster, at which point thereafter, points in the data are classified into clusters according to the k-means algorithm. Because these seeds are, by their nature, randomly chosen, some (e.g. see James et al. 2013) suggest performing many cluster analysis solutions all with different starting seeds, then selecting the solution that best minimizes the within-cluster variance. This is the criteria that the clustering algorithm seeks to minimize, which we discuss now.

12.4 Minimizing Criteria

Recall that so many statistical techniques seek to minimize or maximize some function of the data. Mathematically, fitting statistical models is usually centered around **optimization** of some function. Least-squares regression, for instance, sought to minimize the sum of squared errors. k-Means cluster analysis seeks to minimize a similar objective function, but this time, it is the within-cluster distance that is minimized. This makes intuitive sense of what we would expect from a cluster algorithm of this kind. That is, if the cluster solution is effective, then it should be creating groups of cases that are relatively **homogeneous** in some sense. In most cases, the clustering algorithm ceases when the minimizing algorithm essentially converges on a solution, in the sense that cluster assignments stop changing. One may also wish to specify in advance the number of iterations desired to define the stopping rule for the algorithm.

It is worth highlighting as well that k-means clustering, barring extreme conditions, will typically "find" groups. By the very function it is seeking to minimize, it is **designed to do so**, analogous to how a least-squares line is designed to minimize the sum of squared errors around a line. And while this may be mathematically impressive, on a substantive or scientific basis it may very well not be, and the solution obtained in k-means may or may not make substantive sense to the researcher. Carefully heed the following warning – **simply because a clustering algorithm generates clusters based on a mathematical definition of distance, it does not necessarily imply that these derived clusters represent something the researcher can immediately or ever interpret in a substantive sense**. As mentioned earlier, whether or not clusters "exist" in a real, ontological sense, is a deeper question outside of the mathematical algorithm. The fact that a mathematical algorithm is able to partition points into groups is not the final issue. To the scientist, it is only the beginning.

> By their nature, clustering algorithms will typically find grouping structures in a set of data. However, this does not necessarily imply that such found structures are substantively meaningful. That is, the fact that a cluster program is able to minimize distance criteria does not necessarily imply a meaningful structure to the data.

12.5   Example of *k*-Means Clustering in R

We demonstrate the k-means clustering algorithm in R on the iris data. The iris data is especially suited for a demonstration of the clustering algorithm because, quite simply, we already know in advance one way in which the groups are separated, and that is by **species** of flower. Pause and consider very carefully what we just stated. We repeat it, because it is pedagogically important – **We are already aware of one possible cluster solution (by species), so the iris data is well-suited for verifying whether clustering will find the given solution**. In this sense, we are using clustering as a so-called "unsupervised" learning technique, but we already have a priori knowledge of what the technique should "learn" if successful. Of course, it is also entirely possible that the technique "learn" something else from the data, something that does not agree with a species-classification cluster solution, and in conducting the cluster analysis, we should be prepared for that possibility as well.

We first generate the variables on which we wish to cluster the given cases:

```
> library(car)
> attach(iris)
> iris.data <- data.frame(Sepal.Length, Sepal.Width,
Petal.Length, Petal.Width)
```

```
> some(iris.data)
      Sepal.Length Sepal.Width Petal.Length Petal.Width
3           4.7          3.2          1.3          0.2
12          4.8          3.4          1.6          0.2
14          4.3          3.0          1.1          0.1
33          5.2          4.1          1.5          0.1
37          5.5          3.5          1.3          0.2
86          6.0          3.4          4.5          1.6
116         6.4          3.2          5.3          2.3
122         5.6          2.8          4.9          2.0
129         6.4          2.8          5.6          2.1
```

Note that we could have also defined the variables we wished to use by making direct reference to the original iris data frame instead of defining a new one by:

```
> iris.2 <- iris[,-5]
```

What the above code communicates is to define a new object (iris.2) by using the original data frame variables except for the fifth one. That is, the "−5" subtracts the fifth variable in the data, but includes the first 4. The first 4 variables correspond to Sepal.Length, Sepal.Width, Petal.Length, and Petal.Width, while the fifth variable is Species. To give the variable name Species its own name, we can define an object made up of species only:

```
> species <- iris[,5]
```

That is, the [,5] means to include only the 5th variable (counting from left in the data frame), and give it the name "species."

12.5.1 Plotting the Data

A scatterplot of the iris data may be useful here simply to get an initial picture of what the relationships among variables look like. These plots will not tell us what the final cluster solution will look like, but they will help us see initial separation:

```
> pairs(iris.data, col = species,
+ lower.panel = NULL)
```

In the above code, the line `lower.panel = NULL` will remove the lower triangular of the scatterplot matrix, since this will be simply a mirror-image of the upper triangular part of it since it is a symmetric matrix. The resulting plot appears below:

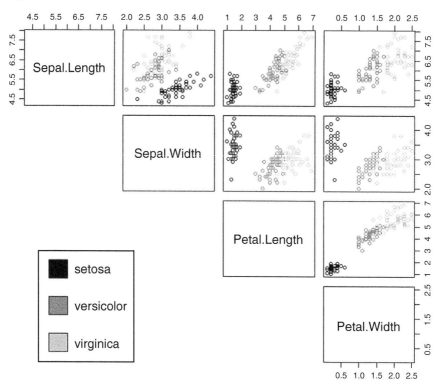

We cannot identify the different species by color since the book is printed in black and white, but if we could, then adding a legend (code adapted from Galili (2019)) to the plot would be useful (which is already included in the above plot):

```
> par(xpd = TRUE)
> legend(x = 0.05, y = 0.4, cex = 2,
+ legend = as.character(levels(species)),
+ fill = unique(species))
> par(xpd = NA)
```

What can we make of the scatterplot matrix? We can see that in virtually all plots, the species setosa is showing quite good separation from species versicolor and virginica taken as a set. For example, in the plot `Sepal.Width` x `Petal.Width`, row 2, column 4, we see that on these two variables, setosa seems to be "on its own island" so to speak, while versicolor and virginica appear much more "together."

We now perform *k*-means:

```
> set.seed(20)
> k.means.fit <-kmeans(iris.data, 3, nstart = 20)
> k.means.fit
```

The 3 above tells R how many clusters we are seeking. Coding centers = 3 (i.e. kmeans(iris.data, centers = 3) would provide the same information:

```
K-means clustering with 3 clusters of sizes 50, 38, 62

Cluster means:
  Sepal.Length Sepal.Width Petal.Length Petal.Width
1     5.006000    3.428000     1.462000    0.246000
2     6.850000    3.073684     5.742105    2.071053
3     5.901613    2.748387     4.393548    1.433871

Clustering vector:
  [1] 1 1 1 1 1 1 1 1 1 1 1 1 1 1 1 1 1 1 1 1 1 1 1 1 1 1 1 1 1 1 1 1 1 1 1 1 1
 1 1 1 1 1 1 1 1 1 1
 [38] 1 1 1 1 1 1 1 1 1 1 1 1 1 3 3 2 3 3 3 3 3 3 3 3 3 3 3 3 3
 3 3 3 3 3 3 3 3 3 3
 [75] 3 3 3 2 3 3 3 3 3 3 3 3 3 3 3 3 3 3 3 3 3 3 3 3 3 3 3 2 3
 2 2 2 2 3 2 2 2 2
[112] 2 2 3 3 2 2 2 2 3 2 3 2 3 2 2 3 3 2 2 2 2 3 2 2 2 2 3
 2 2 2 3 2 2 2 3 2
[149] 2 3
Within cluster sum of squares by cluster:
[1]  15.15100 23.87947 39.82097
 (between_SS / total_SS =  88.4 %)
```

We summarize the above output:

- R reports the cluster means (i.e. "centroids") for each of the input variables. For example, Sepal Length has a mean centroid of 5.006 in cluster 1, a centroid of 6.85 in cluster 2, and a centroid of 5.90 in cluster 3.
- Clustering vector: is the classification of each case in the original data file into the corresponding cluster. For instance, case 1 in the data, which was a case of species setosa, was correctly classified into cluster 1, which is the setosa cluster. Case 2 was also correctly classified into the setosa cluster. Notice that if you count up the values of 1, there is a total of 50:

```
[1]  1 1 1 1 1 1 1 1 1 1 1 1 1 1 1 1 1 1 1 1 1 1 1 1 1 1 1 1 1 1 1 1 1 1 1 1
1 1 1 1 1 1 1 1
[38]  1 1 1 1 1 1 1 1 1 1 1 1 1 1
```

That is, each of the above classifications corresponds to the first cases in the data. What are the first cases in the data? Let's look:

```
> species
  [1] setosa     setosa     setosa     setosa     setosa     setosa
  [7] setosa     setosa     setosa     setosa     setosa     setosa
 [13] setosa     setosa     setosa     setosa     setosa     setosa
 [19] setosa     setosa     setosa     setosa     setosa     setosa
 [25] setosa     setosa     setosa     setosa     setosa     setosa
 [31] setosa     setosa     setosa     setosa     setosa     setosa
 [37] setosa     setosa     setosa     setosa     setosa     setosa
 [43] setosa     setosa     setosa     setosa     setosa     setosa
 [49] setosa     setosa     versicolor versicolor versicolor versicolor
 [55] versicolor versicolor versicolor versicolor versicolor versicolor
 [61] versicolor versicolor versicolor versicolor versicolor versicolor
 [67] versicolor versicolor versicolor versicolor versicolor versicolor
 [73] versicolor versicolor versicolor versicolor versicolor versicolor
 [79] versicolor versicolor versicolor versicolor versicolor versicolor
 [85] versicolor versicolor versicolor versicolor versicolor versicolor
 [91] versicolor versicolor versicolor versicolor versicolor versicolor
 [97] versicolor versicolor versicolor versicolor virginica  virginica
[103] virginica  virginica  virginica  virginica  virginica  virginica
[109] virginica  virginica  virginica  virginica  virginica  virginica
[115] virginica  virginica  virginica  virginica  virginica  virginica
[121] virginica  virginica  virginica  virginica  virginica  virginica
[127] virginica  virginica  virginica  virginica  virginica  virginica
[133] virginica  virginica  virginica  virginica  virginica  virginica
[139] virginica  virginica  virginica  virginica  virginica  virginica
[145] virginica  virginica  virginica  virginica  virginica  virginica
Levels: setosa versicolor virginica
```

We can see from the above that the first 50 cases are indeed setosa. What does this mean for our cluster analysis then? It means the clustering algorithm did a perfect job at classifying the setosa species. Notice, however, that beyond the values of 1 in the clustering vector, the pattern is not quite as perfect:

```
3 3 2 3 3 3 3 3 3 3 3 3 3 3 3 3 3 3 3 3 3 3 3 3
[75] 3 3 3 2 3 3 3 3 3 3 3 3 3 3 3 3 3 3 3 3 3 3 3 3 3
```

Above are enlarged and bolded the two cases that were classified into cluster 2, when they should have been classified into cluster 3. Again, remember, in performing a cluster analysis, we usually do not yet know what the actual clusters are.

If we did, we would be better off doing a **discriminant analysis**. Consider this last remark carefully. In a very strong sense, ANOVA and discriminant analysis can be considered more "evolved" tools of the scientist in which he or she already knows or hypothesizes the group classification, or what machine learning practitioners would call **supervised learning**. It is "supervised" because the user already has designated the groupings, whereas in unsupervised methods the groupings are still yet unknown. However, for the iris data, we are at an advantage because we already have a classifying or clustering scheme in mind, that of **species**. That is, we are able to assess the goodness of the clustering almost akin to if we were doing a discriminant analysis because we know the actual clustering result. We can generate the **confusion matrix** in R:

```
> table(species, k.means.fit$cluster)

species       1  2  3
  setosa     50  0  0
  versicolor  0  2 48
  virginica   0 36 14
```

As we noted already, `setosa` did great, all 50 cases correctly classified. The other two clusters did not fare quite as well.

12.6 Hierarchical Cluster Analysis

We saw that in k-means clustering, the user or researcher specified the number of clusters in advance, and then the clustering algorithm proceeded to assign cases closest to cluster centroids at each step of the procedure. What is most key to k-means is the advance specification of the number of clusters by the user, similar in spirit somewhat to exploratory factor analysis where the researcher specifies in advance the number of factors he or she expects (or hopes!) to find in the data.

A competing alternative to k-means and similar methods is the class of clustering approaches that are **hierarchical** in nature, also sometimes known as **agglomerative**. In hierarchical clustering, though the researcher may have an idea of what he or she will find in the data regarding the number of clusters that exist, this is not specified in advance of the procedure. In hierarchical approaches, a user chooses from a number of methods for fusing (i.e. joining) cases together. Such methods include:

- single linkage
- complete linkage
- average linkage

In **single linkage**, the clustering algorithm first merges the two cases with the smallest distance. Then, the algorithm measures the distance of this newly created cluster with remaining cases in the data, then fuses with that case having the smallest distance with this newly formed cluster. This fusing of cases at each step of the hierarchy is represented pictorially in what is known as a dendrogram, such as the following:

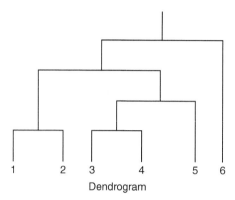

Dendrogram

The dendrogram shows at each step where objects or cases were fused. For instance, in the above dendrogram, we see that objects 1 and 2 were fused, which were then fused to objects 5 and (3,4). The dendrogram provides a kind of "history" of the fusing process, and lets the user immediately see the mathematical similarities of objects with others at different stages of the clustering process. We can summarize more details regarding the above picture of the clustering process:

- At stage 1, (1,2) was formed, as well as (3,4).
- At stage 2, (3,4) was joined with (5).
- At stage 3, (1,2) was fused with [(3,4), (5)].
- At stage 4, (6) was fused with the other clusters at the final stage.

Hierarchical clustering seems simple enough, until we consider the fact that different linkage methods will often yield different clustering solutions. We have already featured single linkage in our example of the dendrogram. However, we also mentioned that complete and average linkage are also possibilities. Whereas in simple linkage it is the smallest of the dissimilarities or distances that is recorded and used in the fusing, in **complete linkage**, it is the largest of the dissimilarities that are computed, the smallest of which are then merged. In **average linkage**, it is the average of the dissimilarities that defines distance. **Centroid linkage** is yet another possibility (for details, see James et al. 2013).

As mentioned, the method of linkage used can have a fairly drastic effect on the ensuing clustering solution, yet there is no agreed upon linkage to use in all circumstances. The overwhelming recommendation by clusters analysis theorists is to employ several methods to see if there is a sense of triangulation among them regarding the number of clusters present. If there is, then this, coupled with researcher expertise in the substantive area, can be used as substantive evidence for the "existence" of clusters.

In hierarchical clustering, there are a variety of linkage options available to the cluster analyst, which include single, complete, and average linkage. Centroid linkage is another possibility. While different linkage methods can yield different clustering solutions, if multiple linkages more or less triangulate on a cluster solution, one might use that as support for the number of clusters existing in the data.

As an example of the different linkage methods put into use, we again feature the iris data, but this time request both single linkage and average linkage solutions. We first build the distance matrix via dist(), then generate the cluster solutions. The dist() function is needed to compute the proximities between all observations in the data file. We will compute Euclidean distances:

```
> d <- dist(iris.data, method = "euclidean")
```

We actually do not need the object d other than for using it in the cluster solution, so calling the object to observe what it produced is not necessary. However, for demonstration, if you did call the object, you would receive for the iris data very (very!) long output containing distances between all observations. For example, we print a few of the cases below:

```
> d
          1         2         3         4         5         6         7         8         9
2   0.5385165
3   0.5099020 0.3000000
4   0.6480741 0.3316625 0.2449490
5   0.1414214 0.6082763 0.5099020 0.6480741
6   0.6164414 1.0908712 1.0862780 1.1661904 0.6164414
7   0.5196152 0.5099020 0.2645751 0.3316625 0.4582576 0.9949874
8   0.1732051 0.4242641 0.4123106 0.5000000 0.2236068 0.7000000 0.4242641
9   0.9219544 0.5099020 0.4358899 0.3000000 0.9219544 1.4594520 0.5477226 0.7874008
```

Above we see that the distance between observations 1 and 2 is equal to 0.5385165, the distance between observations 1 and 3 is equal to 0.5099020, and so on. We now proceed with the two cluster solutions, using single and average linkage, and obtain their respective dendrograms:

```
> clust.single <- hclust(d, method = "single")
> clust.average <- hclust(d, method = "average")
> plot(clust.single)
> plot(clust.average)
```

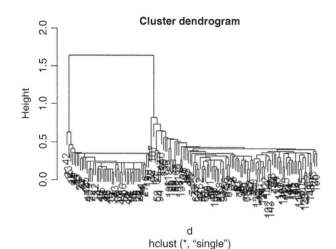

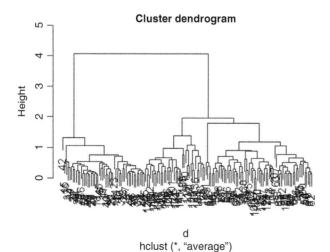

As can be seen from the above, even on a visual basis alone, the different linkage methods suggest quite disparate solutions. Hence, drawing conclusions from a cluster analysis is not at all easy. It is very much an art. While it would appear that two clusters emerge overall from both solutions, several lower-order cluster solutions may exist as well, especially if average linkage is used.

Before conducting a cluster analysis, whether using *k*-means or a hierarchical approach, users need to decide whether or not to **standardize** observations. As recommended by Everitt et al. (2001), in cases where variables subjected to the cluster analysis are not measured in the same units, one option is to standardize variables to have mean zero and unit variance (i.e. transform to z-scores), thus allowing variables to be on par with each other variance-wise before the clustering takes place. Hence, as was the case for principal component analysis, users may choose to standardize their data prior to conducting the analysis if variables subjected to the procedure are not very commensurate in terms of their variances. However, as was the case for PCA, simply because you can standardize does not mean you always should, and many specialists on the subject perform cluster analyses on raw data. If in doubt, run both solutions and compare findings, especially if variables in raw score format are wildly different in terms of variances. There is absolutely nothing wrong with trying out different solutions to see what you get.

12.7 Why Clustering Is Inherently Subjective

While cluster analysis utilizes well-known and mathematically rigorous clustering algorithms, there is, surprisingly, no agreed upon and clear set of rules regarding which measure of proximity to use, the number of clusters to select (in *k*-means), or which type of linkage is "best" in some respect for a given analysis. The guideline given by James et al. (2013) is not encouraging to those who believe cluster analysis will result in finding "true" clusters:

> Each of these decisions can have a strong impact on the results obtained. In practice, we try several different choices, and look for the one with the most useful or interpretable solution. With these methods, there is no single right answer – any solution that exposes some interesting aspects of the data should be considered. (p. 400)

Students studying cluster analysis for the first time often find the above advice somewhat disconcerting if not utterly sobering. That is, they first come to the algorithm with an expectation that it will provide in some sense a **definitive and unique solution**, only to find later that the solution provided is usually one of many possible solutions, not unlike the realities in exploratory factor analysis. **If what you see depends so much on the methodology you use to doing the visualizing, then rigorous evaluations of the lenses by which the seeing is performed seems logical and mandatory**. The problem is that on a mathematical level, the different clustering algorithms can be said to be more or less similar in terms of mathematical rigor. Hence, one cannot usually decide on the "correct" clustering algorithm on this basis alone.

Overall then, one must be cautious with interpretations of cluster analysis. If the methodology one uses is a key determinant in what one "finds," then the finding loses its sense of objectivity. Analogous to factor analysis, the "existence" of clusters often depends on choices made by the researcher, and most of these choices are subjective in nature. What is more, the clustering solution obtained can drastically change when even a small subset of observations are removed at random. Hence, "existence" of clusters could feasibly change depending on simply removing a few observations! This is not unlike factor analysis, in which the removal of a few variables from the factor solution could have rather drastic effects on the solution. Incidentally, this is why some statisticians eventually find their way into the philosophy of statistics, because only by considering the philosophical underpinnings of all of these issues is there a hope for a solution to these problems. Cluster analysis, the technique itself, quite absurdly cannot even address the bigger question for certain of "What is a cluster?" (Izenman, 2008). Only a philosophical inquiry into cluster analysis can even hope to do that. Cluster analysis, the technique, can only offer up a variety of mathematically rigorous ways of defining the nature of a cluster, but no more. Mathematics is truly a rigorous self-contained logical system, but deeper questions about statistical methodology, by necessity, often find their way into philosophical pursuits.

Having said the above, there still have been attempts to provide quantitative validation for cluster solutions in terms of their pragmatic utility. For details, the **fpc** package offers some possibilities and should be consulted.

Exercises

1 Describe the purpose of cluster analysis, and how it might prove useful to a researcher.

2 Discuss the idea of "naturally occurring" clusters. Does a clustering algorithm guarantee such a thing? Why or why not?

3 In the example in the chapter on height and weight, why is it the case that "age" may not be the definitive clustering variable in the data analysis? That

is, why is it the case that the clustering algorithm did not "confirm" that age was the clustering variable?

4 Define the difference between a supervised versus unsupervised learning technique, and why cluster analysis is generally considered "unsupervised." Why would ANOVA generally be considered "supervised?" (i.e. if we had to lump it into one of the two classes)

5 Consider the vectors **z** and **w** with elements (5,10) and (10,20), respectively. Compute the Euclidean distance between these vectors.

6 What is the difference between a k-means and a hierarchical approach to clustering? Further, why is it that the number of clusters in the final solution can be considered arbitrary in each procedure?

7 Consider the USArrests data on which a principal component analysis was performed in a previous chapter. Perform a cluster analysis on this data, and briefly summarize your findings. Use Euclidean distance as your metric.

8 Why is computing a confusion matrix for the analysis in #7 not really applicable as it was when computing one for the iris data example in the chapter? Explain.

9 Consider once more the dendrogram generated in this chapter:

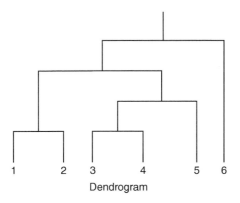

Dendrogram

Is there evidence from the diagram that element 6 is closer to element 5, than is element 3 to 5? Discuss.

10 Why is the determination of the number of clusters inherently subjective? Discuss.

13

Nonparametric Tests

LEARNING OBJECTIVES

- Understand the difference between parametric and nonparametric tests, and when to use them.
- How to run the Mann–Whitney U test (also known as the Wilcoxon rank-sum test) as a nonparametric alternative to the independent-samples t-test.
- How to run the Kruskal–Wallis test as a nonparametric alternative to between-subjects ANOVA.
- How to run the Wilcoxon signed-rank test as a nonparametric alternative to the paired-samples t-test, and the Friedman test as a nonparametric alternative to repeated measures.
- How to run and interpret the Sign test.

In this chapter, we conclude the book with a very brief survey and overview of a few of the more common **nonparametric tests** and demonstrate their use in R. Most of the statistical tests encountered in this book thus far have in one way or another made assumptions about the populations on which the sample data were drawn. For example, when conducting ANOVA we assumed such things as normality of distributions as well as equality of population variances. Many times, however, we may know in advance that our data were not drawn from "well-behaved" populations, and other times it may be essentially impossible to verify assumptions because sample sizes are exceedingly small. It is very difficult to verify assumptions such as normality on samples of size 10, for instance. In these cases and others, such as when data are naturally in the form of **ranks**, nonparametric alternatives may be useful. In some research situations, nonparametric tests are simply the preferred choice, are relatively simple to conduct and interpret, and are able to address simple research questions effectively.

Univariate, Bivariate, and Multivariate Statistics Using R: Quantitative Tools for Data Analysis and Data Science, First Edition. Daniel J. Denis.
© 2020 John Wiley & Sons, Inc. Published 2020 by John Wiley & Sons, Inc.

However, it should be noted from the outset that under most circumstances, parametric tests are still typically preferred over nonparametric ones if assumptions are met. If assumptions are not met, or doubtful, or sample size is very small, nonparametric tests are viable alternatives. It needs to be emphasized as well that there is absolutely nothing wrong in conducting a parametric test and a nonparametric test on the same data to compare outcomes if there is doubt that assumptions are likely not satisfied. Though parametric tests are usually more powerful than nonparametric ones if assumptions are met, a comparison of both approaches on the same data sometimes can prove fruitful in trying to decide whether an effect did or did not occur in one's data.

13.1 Mann–Whitney *U* Test

The Mann–Whitney *U* test is a viable nonparametric alternative to the independent-samples *t*-test, requiring two independent samples on some dependent variable. The Mann–Whitney evaluates the null that both samples were drawn from the same population, but instead of on a continuous measure such as with the original *t*-test, the Mann–Whitney bases its computations on the ranks of the data. Data for the Mann–Whitney should be at least at the **ordinal level**, since if the data are nominal, it becomes impossible to successfully rank them.

We demonstrate the test using the `grades.data` data featured earlier in the book where we conducted the ordinary parametric *t*-test to evaluate the null hypothesis that population means were equal. Recall the grades data:

```
> grades.data <- read.table("grade_data.txt", header = T)
> grades.data
   grade studytime
1      0        30
2      0        25
3      0        59
4      0        42
5      0        31
6      1       140
7      1        90
8      1        95
9      1       170
10     1       120
```

We conduct the test using the `wilcox.test()` function, where we specify `studytime` as a function of `grade`:

```
> attach(grades.data)
> f.grade <- factor(grade)
> mw.test <- wilcox.test(studytime ~ f.grade)
> mw.test

        Wilcoxon rank sum test

data:  studytime by f.grade
W = 0, p-value = 0.007937
alternative hypothesis: true location shift is not equal to 0
```

We summarize the results of the test:

- The null hypothesis for the test is that the distribution of `studytime` is the same across categories of `grade`. The value for W is equal to 0, with a p-value equal to 0.007937. Since this value is less than a cut-off value such as 0.05 (i.e. $p < 0.05$), we reject the null hypothesis and conclude that the distribution of `studytime` is not the same across categories of `grade`.
- R states the alternative hypothesis as `true location shift is not equal to 0` which is more or less the same as concluding the distributions are not the same across grade levels. For more discussion about the alternative hypothesis in this test, see Howell (2002).

13.2   Kruskal–Wallis Test

In the case where we have multiple independent samples, and where we would ordinarily perform a between-subjects ANOVA, the **Kruskal–Wallis** test is the designated nonparametric alternative test. As was the case for the Mann–Whitney U test, we assume the data are at least at the **ordinal level**. We demonstrate this test on the achievement data using the `kruskal.test()` function:

```
> achiev <- read.table("achiev.txt", header = T)
> head(achiev)
  ac teach text
1 70     1    1
2 67     1    1
3 65     1    1
4 75     1    2
5 76     1    2
6 73     1    2
```

```
> attach(achiev)
> f.teach <- factor(teach)
> kruskal.test(ac ~ f.teach)

          Kruskal-Wallis rank sum test

data:   ac by f.teach
Kruskal-Wallis chi-squared = 16.267, df = 3, p-value =
0.0009999
```

We summarize the results of the test:

- The null hypothesis for the test is that the independent samples arose from the same population.
- The test is evaluated by a chi-squared statistic, equal to 16.267 on 3 degrees of freedom, yielding a *p*-value of 0.0009999. Since the obtained *p*-value is less than some preset level such as 0.05 or 0.01, we reject the null hypothesis and conclude that the samples likely did not arise from the same population. Rather, they likely arose from different populations. That is, we conclude that the distributions of achievement are not the same across teachers.

Recall that when we performed the ANOVA on these data in an earlier chapter, after we rejected the overall null hypothesis of equal population means, we wished to conduct post hoc analyses to learn of where the means were different. We can follow up the nonparametric Kruskal–Wallis test using the **Tukey and Kramer** (**Nemenyi**) test in R:

```
> library(PMCMR)
> posthoc.kruskal.nemenyi.test(ac, teach, method = "Tukey")

      Pairwise comparisons using Tukey and Kramer (Nemenyi) test
            with Tukey-Dist approximation for independent samples

data:   ac and f.teach

    1       2       3
2 0.9658 -       -
3 0.3054 0.5849 -
4 0.0014 0.0074 0.2117

P value adjustment method: none
```

What are listed in the table are corresponding *p*-values for comparisons between teachers. We unpack a few comparisons:

- The *p*-value associated with the comparison between teacher 1 and teacher 2 is in the top left portion of the table, equal to 0.9658, and therefore, the difference is not statistically significant. Differences between 1 and 3, 2 and 3 are likewise not statistically significant yielding *p*-values of 0.3054 and 0.5849, respectively.
- Comparisons between 1 versus 4, and 2 versus 4 yield *p*-values of 0.0014 and 0.0074, both statistically significant at $p < 0.05$, while the comparison between 3 and 4 has associated with it a *p*-value of 0.2117, not statistically significant.

13.3 Nonparametric Test for Paired Comparisons and Repeated Measures

13.3.1 Wilcoxon Signed-Rank Test and Friedman Test

A nonparametric alternative to a paired-samples *t*-test is the **Wilcoxon signed-rank test**. Recall the paired-samples *t*-test featured data that were sampled in pairs, and thus the covariance between columns was expected to be other than zero as we would expect in an independent-samples *t*-test. The Wilcoxon signed-rank test considers the magnitudes of differences between pairmates giving more emphasis to pairs that demonstrate larger rather than smaller differences, and then ranking these differences (Howell 2002).

Recall that we used the learning data to show how the paired-samples *t*-test worked in a previous chapter. We will again use it to demonstrate the Wilcoxon signed-rank test:

```
> learn <- read.table("learning.txt", header = T)
> learn
  rat trial time
1   1     1 10.0
2   1     2  8.2
3   1     3  5.3
4   2     1 12.1
5   2     2 11.2
6   2     3  9.1
7   3     1  9.2
8   3     2  8.1
```

9	3	3	4.6
10	4	1	11.6
11	4	2	10.5
12	4	3	8.1
13	5	1	8.3
14	5	2	7.6
15	5	3	5.5
16	6	1	10.5
17	6	2	9.5
18	6	3	8.1

Though there are 3 trials in the learning data, we utilize only the first 2 to demonstrate the test. We need to create two new vectors for both trial 1 and trial 2, as follows:

```
> trial.1 <- c(10, 12.1, 9.2, 11.6, 8.3, 10.5)
> trial.2 <- c(8.2, 11.2, 8.1, 10.5, 7.6, 9.5)
> wilcox.test(trial.1, trial.2, paired = TRUE)
```

Note above that trial.1 consists of only data points for that trial. That is, the first observation is 10, the second is 12.1, etc., for successive rats. We do the same for trial.2, then in the model statement, include the paired=TRUE. The output of the test follows:

```
        Wilcoxon signed rank test with continuity correction

data:   trial.1 and trial.2
V = 21, p-value = 0.03552
alternative hypothesis: true location shift is not equal to 0

Warning message:
In wilcox.test.default(trial.1, trial.2, paired = TRUE) :
  cannot compute exact p-value with ties
```

We summarize the results of the test:

- The null hypothesis for the test is that both trials were drawn from identical populations. The alternative hypothesis is that they were not, or, as R puts it, "true location shift is not equal to 0."
- The obtained p-value for the test is 0.03552, which is less than 0.05, so we reject the null hypothesis and infer the alternative hypothesis. That is, we have evidence to suggest that the trials were not drawn from the same population.

Now, suppose we would like to conduct a test on all 3 trials. For this, we would use the **Friedman test**. For this test, we will leave our data in the original format, and specify the function as `time ~ trial|rat`, which reads that "time is a function of trial, and measurements on trials are contained (or 'blocked' or 'nested') within rat":

```
> attach(learn)
> fried.test <- friedman.test(time ~ trial|rat)
> fried.test

        Friedman rank sum test

data:   time and trial and rat
Friedman chi-squared = 12, df = 2, p-value = 0.002479
```

The *p*-value for the test is equal to 0.002479, which counts as evidence against the null hypothesis and suggests there is a difference, somewhere, between trials. To know where those differences may be, we can follow up with Wilcoxon signed-rank tests to conduct further pairwise difference tests. We will conduct comparisons between trials 1 and 2, 1 and 3, 2 and 3 (you will need to generate the `trial.3` vector below):

```
> wilcox.test(trial.1, trial.2, paired = TRUE)

        Wilcoxon signed rank test with continuity correction

data:   trial.1 and trial.2
V = 21, p-value = 0.03552
alternative hypothesis: true location shift is not equal
to 0

> trial.3 <- c(5.3, 9.1, 4.6, 8.1, 5.5, 8.1)

> wilcox.test(trial.1, trial.3, paired = TRUE)

        Wilcoxon signed rank test

data:   trial.1 and trial.3
V = 21, p-value = 0.03125
alternative hypothesis: true location shift is not equal to 0

> wilcox.test(trial.2, trial.3, paired = TRUE)

        Wilcoxon signed rank test with continuity correction
```

```
data:   trial.2 and trial.3
V = 21, p-value = 0.03552
alternative hypothesis: true location shift is not equal to 0
```

We summarize the results of the Friedman test and follow up Wilcoxon signed-rank tests:

- The null hypothesis is that the three samples were drawn from the same population. The alternative hypothesis is that they were not.
- The obtained p-value is equal to 0.002479, which is less than 0.05 (or even 0.01), leading us to reject the null hypothesis and infer the alternative hypothesis that the samples were not drawn from the same population.
- Follow-up of Wilcoxon signed-rank tests revealed that pairwise differences exist between all trials ($p < 0.05$). One might also wish to use a more conservative significance level for these tests (e.g. such as 0.01) as to limit the potential inflation of the type I error rate.

13.4   Sign Test

The sign test is a nonparametric test in which we are interested in assessing differences between two groups of paired observations, but unlike the paired-samples t-test, we are not interested or able to measure **magnitudes** of differences between pairs. That is, we either do not have data that are measurable on a continuous or pseudo-continuous scale, or we are not interested in measuring differences in terms of magnitudes. Rather, we are only interested in the **sign** of the difference, that is, whether one pairmate scored higher or lower than the other pairmate and recording this as a "+" or "−." Because of its nature, the sign test actually uses the **binomial distribution**, and as such, is nothing more than a binomial test.

As an example of the sign test, consider the following data on fictitious marital happiness scores for husband and wife pairs, where 10 is "most happy" with the marriage, and 1 is "least happy" with the marriage. Each pair below consists of a husband and wife on which happiness scores were recorded. For example, for the first pair, the husband scored a value of 2 and his wife a value of 3. Since we are subtracting the wife's score from that of the husband, the difference is $2 - 3 = -1$. However, for the statistical test, we don't care about the magnitude of the difference. All we care about is the **sign**, and so a "−" is recorded in the column **Sign (H–W)** which is short for "the sign obtained from husband's score minus wife's score."

Pair	Husband	Wife	Sign (H–W)
1	2	3	−
2	8	7	+
3	5	4	+
4	6	3	+
5	7	9	−
6	10	9	+
7	9	10	−
8	1	3	−
9	4	3	+
10	5	6	−

We can see from the distribution of signs (rightmost column) that there are 5 "+" and 5 "−" signs. It stands that if there were no differences between husbands and wives on marital satisfaction scores, in the sense of + versus −, then we would expect the **number of signs to be equal**. This is exactly what we're seeing with our data. On the other hand, if there were differences between husbands and wives, we'd expect the number of + (or −) signs to be **unequal**. It's easy to see then that this is a case where the binomial distribution can be used to evaluate the number of positives against negatives.

Recall that the binomial distribution models the number of successes versus failures, and is easily implemented in R. For convenience, we will count the number of + signs as "successes" and the number of − signs as "failures" (though either way would work). In R, we compute a significance test for observing 5 successes out of 10 trials:

```
> binom.test(5, 10)

        Exact binomial test

data:  5 and 10
number of successes = 5, number of trials = 10, p-value = 1
alternative hypothesis: true probability of success is not
equal to 0.5
95 percent confidence interval:
 0.187086 0.812914
sample estimates:
probability of success
                  0.5
```

We summarize the results of the test:

- The null hypothesis is that, overall, there is no difference between marital satisfaction between husbands and wives. The alternative hypothesis is that, overall, there is. Interpreted in the context of a binomial distribution, the null is that the probability of success is equal to 0.5, while the alternative is that the probability of success is not equal to 0.5.
- Five of the difference scores are positive, and five are negative. The exact test yields a *p*-value of 1.0. Hence, we do not reject the null hypothesis. That is, we do not have evidence to conclude that overall, a difference exists between marital satisfaction between husbands and wives.
- R reports a 95% confidence interval indicating plausible values for the probability of success.

Exercises

1   Describe a research situation where a nonparametric test may be preferred over a parametric one.

2   Earlier in the book we featured the Spearman rho correlation coefficient. Discuss why this may be considered a nonparametric statistic. That is, what about it makes it nonparametric?

3   Consider the following data on vectors **x** and **y**, where **y** is a grouping variable:

x: 0, 4, 8, 10, 15, 20, 21, 22, 25, 30.
y: 0, 0, 0, 0, 0, 1, 1, 1, 1, 1.

Perform a Mann–Whitney *U* test in R, and briefly summarize your results.

4   Perform a univariate *t*-test on the data in #3, and compare your results to that of the nonparametric test. How do they compare? Has your conclusion on the null changed at all?

5   Perform a Kruskal–Wallis test on the achievement data, this time with text as the independent variable (and achievement scores, as before, as the response). Summarize your findings.

6   Does it make sense to perform a post hoc test on the data in #5? Why or why not?

7 In Chapter 1 of this book, we featured a discussion of the coin-flip example to illustrate how hypothesis-testing works. How would the sign test of this current chapter be applicable to the coin-flip example? Discuss.

8 Your colleague conducts a Mann–Whitney U test, and obtains a p-value of 0.051, concluding "no difference." Comment on your colleague's conclusion. Do you agree with it? Why or why not?

9 Conduct a one-way repeated measures analysis on the learn data of this chapter on which the Friedman test was conducted. Compare p-values based on the parametric test to that of the nonparametric. To what extent are they different from one another? If they are different, why may this be so?

10 Earlier in the book, we briefly made mention of a technique called "bootstrapping." How does that procedure relate to methods covered in this chapter? Explain the link by doing a brief online search of the technique.

References

Agresti, A. (2002). *Categorical Data Analysis.* New York: Wiley.

Baron, R. M. & Kenny, D. A. (1986). The moderator–mediator variable distinction in social psychological research: Conceptual, strategic, and statistical considerations. *Journal of Personality and Social Psychology, 51,* 1173–1182.

Cohen, J. (1988). *Statistical Power Analysis for the Behavioral Sciences.* New York: Routledge.

Cohen, J., Cohen, P., West, S. G., & Aiken, L. S. (2002). *Applied Multiple Regression/ Correlation Analysis for the Behavioral Sciences.* New Jersey: Lawrence Erlbaum Associates.

Crawley, M. J. (2013). *The R Book.* New York: Wiley.

Dalgaard, P. (2008). *Introductory Statistics with R.* New York: Springer.

Denis, D. (2016). *Applied Univariate, Bivariate, and Multivariate Statistics.* New York: Wiley.

Denis, D. (2019). *SPSS Data Analysis for Univariate, Bivariate, and Multivariate Statistics.* New York: Wiley.

Everitt, B. & Hothorn, T. (2011). *An Introduction to Applied Multivariate Analysis with R.* New York: Springer.

Everitt, B. S., Landau, S., & Leese, M. (2001). *Cluster Analysis.* New York: Oxford University Press.

Fiedler, K., Schott, M., & Meiser, T. (2011). What mediation analysis can (not) do. *Journal of Experimental Social Psychology, 47(6),* 1231–1236.

Fitzmaurice, G. M. (2011). *Applied Longitudinal Analysis.* New York: Wiley.

Fox, J. (2016). *Applied Regression Analysis & Generalized Linear Models.* New York: Sage.

Galili, T. (2019). Hierarchical cluster analysis on famous data sets – enhanced with the dendextend package. Retrieved on September 2, 2019. https://cran.r-project.org/web/packages/dendextend/vignettes/Cluster_Analysis.html

Univariate, Bivariate, and Multivariate Statistics Using R: Quantitative Tools for Data Analysis and Data Science, First Edition. Daniel J. Denis.
© 2020 John Wiley & Sons, Inc. Published 2020 by John Wiley & Sons, Inc.

Guttag, J. V. (2013). *Introduction to Computation and Programming Using Python.* Cambridge: MIT Press.

Hastie, T., Tibshirani, R., & Friedman, J. (2009). *The Elements of Statistical Learning: Data Mining, Inference, and Prediction.* New York: Springer.

Hays, W. L. (1994). *Statistics.* Fort Worth, TX: Harcourt College Publishers.

Hennig, C. (2015). What are the true clusters? *Pattern Recognition Letters, 64,* 53–62.

Hintze, J. L. & Nelson, R. D. (1998). Violin plots: A box plot-density trace synergism. *The American Statistician, 52(2),* 181–184.

Holtz, Y. (2018). *The R Graph Gallery.* Document Retrieved on September 2, 2019. www.r-graph-gallery.com

Hosmer, D. & Lemeshow, S. (1989). *Applied Logistic Regression.* New York: Wiley.

Howell, D. C. (2002). *Statistical Methods for Psychology.* Pacific Grove, CA: Duxbury Press.

Izenman, A. J. (2008). *Modern Multivariate Statistical Techniques: Regression, Classification, and Manifold Learning.* New York: Springer.

James, G., Witten, D., Hastie, T., & Tibshirani, R. (2013). *An Introduction to Statistical Learning with Applications in R.* New York: Springer.

Johnson, R. A. & Wichern, D. W. (2007). *Applied Multivariate Statistical Analysis.* New Jersey: Pearson Prentice Hall.

Jolliffe, I. T. (2002). *Principal Component Analysis.* New York: Springer.

Kassambara, A. (2017). *Machine Learning Essentials: Practical Guide in R.* STHDA.

Kirk, R. E. (2012). *Experimental Design: Procedures for the Behavioral Sciences.* Pacific Grove, CA: Brooks/Cole Publishing Company.

Kuhn, T. S. (2012). *The Structure of Scientific Revolutions: 50th Anniversary Edition.* Chicago: University of Chicago Press.

MacKinnon, D. P. (2008). *Introduction to Statistical Mediation Analysis.* New York: Lawrence Erlbaum Associates.

MacQueen, J. (1967). Some Methods for Classification and Analysis of Multivariate Observations. In: L. LeCam & J. Neyman (Eds.), *Proceedings of the Fifth Berkeley Symposium on Mathematical Statistics and Probability,* Vol. 1. Berkeley: University of California Press, pp. 281–297.

Matloff, N. (2011). *The Art of R Programming.* San Francisco: No Starch Press.

McCullagh, P. & Nelder, J. A. (1990). *Generalized Linear Models.* New York: Chapman & Hall.

Mulaik, S. A. (2009). *The Foundations of Factor Analysis.* New York: McGraw-Hill.

Olson, C. L. (1976). On choosing a test statistic in multivariate analysis of variance. *Psychological Bulletin, 83,* 579–586.

Petrocelli, J. V. (2003). Hierarchical multiple regression in counseling research: Common problems and possible remedies. *Measurement and Evaluation in Counseling and Development, 36,* 9–22.

Rencher, A. C. (1998). *Multivariate Statistical Inference and Applications*. New York: Wiley.

Rencher, A. C. & Christensen, W. F. (2012). *Methods of Multivariate Analysis*. New York: Wiley.

Schumacker, R. E. (2016). *Using R with Multivariate Statistics*. New York: Sage.

Searle, S. R., Casella, G., & McCulloch, C. E. (2006). *Variance Components*. New York: Wiley.

Stigler, S. M. (1986). *The History of Statistics: The Measurement of Uncertainty Before 1900*. London: Belknap Press.

Stoltzfus, J. C. (2011). Logistic regression: A brief primer. *Academic Emergency Medicine*, *18*, 1099–1104.

Tabachnick, B. G. & Fidell, L. S. (2001). *Using Multivariate Statistics*. New York: Pearson Education.

Warner, R. M. (2013). *Applied Statistics: From Bivariate Through Multivariate Techniques*. London: Sage.

Wickham, H. (2009). *ggplot2: Elegant Graphics for Data Analysis*. Springer: New York.

Wickham, H. (2014). *Advanced R*. New York: Chapman & Hall/CRC The R Series.

Yuan, K. & Bentler, P. (2000). Inferences on correlation coefficients in some classes of nonnormal distributions. *Journal of Multivariate Analysis*, *72(2), 230–248*.

Index

Univariate, Bivariate, and Multivariate Statistics Using R: Quantitative Tools for Data Analysis and Data Science, First Edition. Daniel J. Denis.
© 2020 John Wiley & Sons, Inc. Published 2020 by John Wiley & Sons, Inc.